Mathematics
Teaching and Learning

Mathematics Teaching and Learning

Jon L. Higgins

The Ohio State University

Charles A. Jones Publishing Company

Worthington, Ohio

1   2   3   4   5   6   7   8   9   10   /   77   76   75   74   73

Library of Congress Catalog Card Number: 70-184675

International Standard Book Number: 0-8396-0013-5

Printed in the United States of America

Preface

Good teaching requires both a practical and a theoretical frame of reference. *Mathematics Teaching and Learning* brings together these two aspects of education. It is a book of things to do: learning experiments to try, evaluate, and revise. Analyzing the data will require the use of simple statistics. All the statistical tools needed are fully explained in this book, which can serve as a starting point for exploration of the vast world of statistics.

Mathematics Teaching and Learning contains readings by leading psychologists in the area of learning theory. They are "position papers" of major conjectures and theories about how we learn. They raise a number of thoughtful, provocative questions. Alternative theories or modifications of earlier theories are presented. Each of these contains useful ideas, but the final learning theory has yet to be discovered. The ways in which children learn mathematics is still an active area of research. *Mathematics Teaching and Learning* reflects this state of activity.

Mathematics Teaching and Learning also contains discussion sections which examine the implications of various learning theories for classroom teaching. The ways in which one can form, write, and classify classroom objectives are developed from a study of intellectual abilities. Lesson planning and organization is measured against both psychological principles and the logical principles inherent in mathematics itself. Lesson presentation is considered by exploring ways in which mathematics can be personalized to achieve maximum student involvement. Even hints for classroom discipline are drawn from psychological principles. These discussion sections represent the final combination of theory and practice.

Mathematics Teaching and Learning is organized into *Units*. Each unit treats one psychological area or theory. The activities which relate to the unit topics are *Modules*. The Investigation Modules are experiments. Major theories or positions are contained in Reading Modules. Applications of these theories to classroom practice comprise the Discussion Modules. Each unit in *Mathematics Teaching and Learning* concludes with suggestions for additional activities. These activities will help the reader expand the study and understanding of the unit's topic. The section *For Related Readings* and *For Related Research* suggest further explorations in the literature. The section *For Microteaching* is designed for practice in teaching applications developed in each unit. Microteaching is a small controlled situation which allows the teacher to use a single teaching technique. The number of students and the length of teaching are both limited to focus on specific teaching skills. The activities can be adapted for use as field experiences in actual school settings.

Mathematics Teaching and Learning uses mathematical examples from the newest mathematics curricula. It is not a substitute for careful study of mathematics content or the methods of presenting specific content, but the psychological frame of reference developed by this book will help make the study of such mathematics content more meaningful and useful to prospective teachers.

Experienced teachers will find this book useful for analyzing their current teaching practices. Most teachers can add examples from their own situations to the teaching examples found in the book. In so doing, they will begin to discover basic principles which can be used to modify old practices and create new strategies to meet new challenges in mathematics teaching.

Advanced students in mathematics education will find the book to be a useful survey of current positions in the psychology of mathematics learning. The readings and activities of the book can serve as springboards for further reading and will suggest research areas for future investigations.

Jon L. Higgins

Contents

The Charles A. Jones Publishing Company

International Series in Education

What does it mean to teach mathematics? What do mathematics teachers actually *do* in the classroom?

Every prospective mathematics teacher is faced with these questions as he prepares to become a teacher. Every inservice teacher faces these questions, too, as he evaluates his teaching in light of that of his colleagues. The questions seem simple, and the answers should be simple, too. To find out what goes on in the mathematics classroom, one should walk in and observe.

Unfortunately, the questions are not quite that simple. Implicit in them is not only a sense of the probable, but a sense of the possible as well. What is the possible range of sequences, strategies, and moves that lead to good mathematics teaching? Good teachers have always sought answers to this question, and master teachers are marked by their continuing search for new techniques and strategies. The study of instruction contains no clear-cut structure to aid the novice — there are many ways to study teaching, and many ways to categorize the instructional act.

One of those ways is to look at the intent of the teaching act. Many teaching acts help the student *explore* mathematics. This kind of teaching assumes that mathematics can be experienced. But it makes even more assumptions about the learner. It assumes that children are naturally curious— that if given opportunities to explore they automatically take advantage of those opportunities. It also assumes that children can discriminate among experiences those underlying patterns and structures we associate with mathematics. Both of these assumptions bear careful investigation.

Other teaching acts serve to *model* mathematical thought and action. Teaching by modeling involves the careful use of both examples and nonexamples. It assumes that children have the ability to generalize from specific instances to broader categories and classifications. The study of this assumption comprises a major branch of educational psychology.

Closely related to modeling is a category of teaching acts that might be called *underlining*. Here the intent of the instruction is to emphasize logical relationships and critical steps in a mathematics sequence. Where modeling often emphasizes variety, underlining focuses upon developing mathematics in an orderly, introspective, and often deductive manner. It assumes that children can understand implications and consequences that follow from specific assumptions or actions. It also assumes that proper order and sequence is an important factor in the learning process.

Yet another category of instruction might be termed *challenging*. This kind of instruction encourages students to supply steps, justifications, and evaluations on their own. It emphasizes that learning is not only passive reception but also active involvement. It assumes that learners can examine parts of a structure and generate the remaining details of the structure.

A final important category of instruction is *practicing*. Building upon a philosophy of "learning by doing," this instruction also seeks to consolidate desirable actions and procedures into definite algorithms. It assumes that learning requires more than a single instance of perception, and that retention depends upon the accumulation of experience.

These five categories — exploring, modeling, underlining, challenging, practicing — only begin to suggest the wide variety of instructional acts necessary in good mathematics teaching. In this book, the primary concern will be to carefully examine the assumptions about learning that underlie these categories. Examination of theories of learning and data that support them provide one way of evaluating instructional techniques that rest on these theories. Learning theories also suggest possible alternate strategies of teaching. Thus, the book is organized so that all major contemporary theories of learning are examined and their implications for mathematics teaching discussed.

Before embarking upon this study of learning, we will briefly examine the implications of the five categories for classroom teaching. In particular, we should examine the adequacy of the categories themselves. Are the categories broad enough to encompass all teacher behavior? Can the categories be used in classroom observations? Can rules be formulated that guide the observer in deciding what specific acts should be included in each category? Can we find additional assumptions about learning that are implicit in the categories?

Variety Within Categories

Within each of these categories, one can observe a wide range of teaching actions. The teacher who uses *exploring* techniques may focus upon establishing a stimulating environment. His classroom may appear to be a laboratory filled with materials and games. When the interaction of each child with materials is emphasized the teacher is not a dominant figure in the classroom. Instead he circulates throughout the room, following the leads of students, supplying the encouragement necessary to maintain levels of discovery. But there is a fine dividing line between discovery and confusion or between freedom and chaos. Successful teaching with exploring techniques demands detailed knowledge of both students and the mathematics topics to be taught.

Modeling introduces students to the best techniques that have proved useful to great mathematicians. The teacher can choose and introduce examples that reflect both the power and beauty of mathematics. But teaching by modeling also requires skill, lest it degenerate to a "follow-the-leader" process where all actions are initiated by the teacher. Innovative teachers also present models by means of supplementary readings, films, and small group discussions.

Underlining usually occurs in an expository format. Yet underlining need not simply be a recapitulation of textbook material or lecture notes. The emphasis of relationships and critical ideas can also be developed through discussions and debate. The most effective underlining sometimes results from simply asking "what would happen if we did it another way?"

Challenging means more than simply questioning. Many questions ask for a mere repetition of material. Questions that challenge are of a higher order than this. The challenging teacher is not afraid to become a devil's advocate or to set up a misleading situation which can be easily torpedoed. A class taught by methods of challenging crackles with both wit and excitement.

Few mathematics classes are complete without *practicing*. Practicing can be as uninspiring as repeated drill. Or it can be games, contests, and applications. Practicing suggests shortcuts, patterns, and mnemonics to aid remembering.

Making Classroom Observations

The art of successful classroom observation depends upon being able to establish a set of references. A typical class of thirty students usually presents a bewildering variety of actions to the novice observer. If an average action (incident) lasts fifteen seconds, observing a class of thirty students over a fifty-minute class period can present the observer with as many as 6000 incidents. Understanding what has happened during this period demands a frame of reference that will help organize the impressions gathered. The five teaching classifications we have discussed provide one such frame of reference.

An actual teacher may act in several different ways in the course of a class period. It may be helpful to record the sequence of categories used to try to describe an over-all strategy that is being used. For example, a teacher may begin a lesson by giving students an opportunity to explore. If interest lags before the desired relationships have been discovered, he may switch to a challenging category to encourage students to look at the problem in different ways. He might then return to an exploring category by assigning different groups to look at different embodiments of the concept. Perhaps this activity will be followed by a period of underlining which reviews separate discoveries and relates them to each other. The teacher may then model a particular technique or algorithm useful in applying the concept, and finally assign a set of exercises for the purpose of practicing. The observer will surely want to examine the category pattern of "exploring-challenging-exploring-underlining-modeling-practicing" to see if the strategy is one which this teacher uses often, or if it is a technique useful only in teaching the particular topic that was observed.

Often it is helpful to observe the relative amounts of time a teacher spends within each category during a class period. Does the teacher spend twenty percent of the time in each category, or is the great majority of each class period spent in the same kind of activity? What kinds of distributions are desirable for the kinds of topics and the ages of the students involved?

Observing specific teaching acts and classifying them successfully depends upon determining the intent of the act. This can often be deduced from the context of the act. Suppose a teacher asks a class to flowchart the steps involved in squaring a binomial. If this activity takes place early in the study of the topic, it is probably an underlining activity, intended by the teacher to reveal and clarify the steps involved in the process. If, on the other hand, students are asked to construct the flowchart after they have already worked several examples, the activity may be one of practicing.

The art of successful teacher observation may ultimately depend upon focusing on students. Does the instructional activity cause the student to explore or only to observe? If the student only observes, then the proper category for the instruction may be modeling. Does a teacher's question cause the student to supply a new idea? If so, the act may be challenging; but if not, it may really be underlining by means of repeating an important fact.

Activity: Three Classroom Observations

There is no substitute for on-the-spot classroom observations. However, the following activity will help in preparing for such observations. The following material is a transcript of short excerpts from mathematics classes at three different grade levels. These excerpts are contained in the film, *Mathematics for Tomorrow,* available from the National Council of Teachers of Mathematics. View the film so that you may see the actions and reactions of both students and teachers. If viewing is not possible, then read the dialogs carefully and imagine what students are doing as each teacher talks. Answer the following questions for each teacher:

1. How would you characterize this teacher's actions? Does he use a variety of categories that seem to form a strategy, or does he stay within one category?
2. What does this teacher expect students to be doing in his class? Do you think students would be active or passive in this classroom?
3. Can you find a word or term which describes the approach which the teacher seems to be taking with his class? What does this teacher seem to believe about the nature of students and the way they learn mathematics?

When you have finished all three transcripts, write a short paragraph listing similarities and differences you observed in the teaching techniques used at the three different grade levels.

(One word of caution — like many areas of education, there are no single "right answers" to these questions. You should consider a range of possibilities and weigh the merits of each. The real learning process in this activity comes when you compare your analysis with the analyses that others have made.)

A First Grade Class Taught by Mrs. Martha Gilman

The first graders are seated around a low table with Mrs. Gilman. Several objects are on the table, including a paper sack, and two large metal cans.

"Let's find out what I have in this sack. There's something you can eat in this sack. What do you suppose I have in this sack? What do you think, Greg?"

"Candy!"

"No, not candy. It's something even better for your teeth than candy. What do you think, Jan?"

"Chocolate!"

"No, I have a whole set of apples in that sack."

"They're real?"

"They're real! Rita asked a good question, 'Are they real?' " (The apples are removed from the sack one by one.) "They're real apples. Now I would like you to just use your eyes and look around our room. See if you can find any other sets of things in our room, and then raise your 'signal' (hand) if you can find other sets of things in our room. Lou Anna?"

"The things on the blackboard?"

"The set of charts on the blackboard? Good! Anybody else find some sets of things in our room? Louis."

(Louis points.)

"You mean the numbers in the yellow circles on the back wall?"
(He shakes his head no.)

"I don't know what you're pointing to, dear. You mean the papers in our room? Yes, the set of papers behind us on the wall. Toby?"

(The dialog continues while students point out more examples of sets. The concept of members of a set is introduced. The attention turns to the three apples on the table and three pieces of chalk.)

"Are the members of this set the same as the members of this set? Are they the same thing?"

"Yes"

"Do I have apples over here?"

"No."

"Are the members the same then?"

"No."

"No. Greg, you keep shaking your head 'yes'. The members are not the same thing they all decided. Why are you shaking your head 'yes', Greg? (No response.) What is the same — is there something the same about these two sets?"

"Yes," (from Jerry).

"Jerry, what's the same about these two sets?"

"They're different, and . . ."

"Yes, we said the members were different. But there's something the same about these two sets. Rita, what's the same?

"Both three."

"How do you know they both have three things, Rita? How could you tell? You told me they each had three members — each set had three members. How could you tell? (No response.) Louis, how could you tell?"

"Because they look the same."

"They look the same! You can just kind of look at those two sets and tell that they both have three members in them. All right, but now I'll ask something that's a little harder. I have a can of erasers, and in my can I have a whole set of erasers, and in this can I have a set of crayons. Now you can't tell if these sets match or not because you can't see inside. How am I going to find out if I have the same amount of members in the set in this can as I have in the set in this can over here. Jan Dougal, how am I going to find out?"

"Open the top."

"Open the top, and . . . ?"

"Pull them out."

"How would I pull them out? How would I find out if this set has the same amount of members as this set. How would I . . . ? Eddie, how would I find out?"

"Match them."

"Match them! Good for you! Eddie, can you do that for us?"

A Seventh Grade Class Taught by Mr. Clayton Ross

The class is looking at a large drawing of a right rectangular prism. Each side has been subdivided into units, and these unit lines have been extended across the faces so that the prism appears to be a stack of cubes. Don is holding the chart, which appears to be one of a set.

"Let's try this one, Don. Judy, what would you say is the length?"

"Three."

"Three. What would you call the width? Lester?"

"Two."

"All right. What would you say is the height . . . Claudette?"

"Two."

"All right, and what do you think is the volume, Clair?"

"Twelve."

"Twelve. Do you agree to that, do you think it's twelve, Candy?"

"Yes, I do."

"Do you think it's twelve? Nancy, what would you say was the volume there? How many cubes in the whole thing?"

"In the whole thing?"

"In the whole thing. In the whole rectangular solid. You can't see all those cubes; some are underneath the stack there."

"Twenty-eight?"

"Twenty-eight, huh? Let's see. Tom, how many pieces does it look like to you in the whole stack?"

"Twelve."

"You think there are twelve there? All right. How about you, Judy, we haven't given you a chance."

"Twelve."

"About twelve. Well, we don't all agree, but we kind of get the idea; many of you seem to think there are about twelve cubes there. Why don't we do it this way? Here are the cubes you people made. (He points to several paper cubes.) Let's stack them up just like in the picture. Let's see. Three along like this . . . . (He stacks the cubes.) Does that look like the picture? Let's see, we've got length three, three cubes along that way. Width two, two cubes along this way; depth two, well we've got height or depth two there. That makes six cubes in the top layer, and six more in the bottom layer, so twelve's right. Thank you, Don. (Don returns to his seat.) We've done this several times now, people; we've done it with a number of pictures; we've got the information tallied on the board. Does anyone see any kind of a rule or relationship between these numbers — we have the length, the width, and the depth — and the number we wrote down for the volume. What do you see, Todd?"

"You take the length number times the width number, and then you multiply that number by the height and that's the volume."

An Eleventh Grade Class Taught by
Mr. Robert E. K. Rourke

"Now this morning I want to discuss the probabilistic model for finding a needle in a haystack. Perhaps I should say a needle in a needle-stack, and here it is. (He walks over to a glass aquarium which has been filled with plastic spools.) This container has a great many machine parts. They all look exactly alike. But the fact of the matter is, they are not all exactly alike. Three or four mixed in here at random have been made by a special precise process that's very costly. Now, I can't find these three or four special items by looking, but I can find them by testing. The test takes time and money. Here's the question: should I test or should I go out and buy new ones? Now in order to answer this question, among other things I'd like to know is this: on the average, how many of these will I have to test before I find the special one? The probabilistic model that I have for solving problems of this type is a very simple one: an ordinary deck of cards. (He picks up a deck of cards.) I use the four aces in the deck for special items. And I use the rest of the deck, the forty-eight nonaces, for nonspecial items. And I'm going to approach this probabilistic model in the usual way, from three points of view. We'll find out from it what we can by guessing; we'll find out what we can by experimenting, and finally we'll apply our theoretical mathematics to the model. Now, I'm going to describe an experiment for you. I want you to follow me closely, and then at the end I want you to guess the outcome of this experiment. I'm going to put one needle in the haystack — one ace in the stack. I'm going to shuffle thoroughly. (He shuffles his deck of cards.) I'm going to count until I find that ace. I'm going to note its position number. And then I'm going to repeat the experiment a hundred times. I have a hundred position numbers. I will average them all and I will get the average position of the ace. Now, there are forty-eight nonaces in a row. Where would you expect on the average to find that ace? Henry? (No response.) Oh come on, guess. The guesses are free! Where would you expect to find it? (No response.) Well, Henry, would you expect to find it near the front on the average? Near the back? Alice, where would you expect to find it?"

"Near the middle?"

"Exactly! Near the middle. There's some symmetry involved in this, and we sort of feel in our bones that that ace is going to divide the forty-eight string into two equal parts, twenty-four in the first part, so somewhere about position twenty-five we expect to find our ace. I add another needle to the haystack. I perform the same experiment another hundred times. Where would the average positions of the aces be this time?"

(The guessing procedure continues for two, three, and four aces in the deck.)

"So much for guessing. Now we want to try the experimental part of our program. I want everyone tonight to do this experiment with the four aces twenty times. Get me twenty position numbers for the first ace. In the whole class that will give us about three hundred position numbers, and we will see if experiment backs up what we found out by guessing. And then tomorrow we'll do the most interesting part of all; we'll apply our theoretical mathematics to this problem."

Summary: Focus and Degree

The purpose of the preceding activity has not been to catalog all the possible practices of mathematics teachers but to attempt to simplify the problem by looking at broad categories of actions. This simpler way of looking at the classroom permits one to look at the assumptions teaching acts imply about how children learn mathematics. Looking at these assumptions shifts the focus from teacher to student — from a teacher's actions to what these actions imply about students. What does the teacher believe about the basic nature of students? Why do students have different abilities in mathematics? How do students learn mathematics?

All teachers have assumptions which they hold as answers to these questions, and which, in turn, influence their teaching of mathematics. In many cases, these assumptions are only implicit. They may be well-formed, but the teacher would find it extremely difficult to verbally explain them consistently and coherently.

Often a teacher's assumptions about children are far too simple. Human beings are extremely complex, and assumptions about students need to be diverse enough to allow for this complexity. A single good teaching practice is not effective if it represents the *only* thing the teacher ever does.

A teacher who lectures (underlines) constantly without ever stopping to question his students to find out their reactions is not effective. A teacher who only asks questions without ever "telling things the way they are" would soon alienate his students just as much as the constant lecturer. A teacher who has students discover mathematical patterns but never stops to discuss what these discoveries mean or how they might be used is hardly teaching. Effective teaching needs teachers who are flexible and who can act in a variety of ways when appropriate. Their assumptions about students, student abilities, and the way students learn mathematics need to be broad enough to encompass a wide variety of situations.

We began by asking what it means to teach mathematics. We looked at teacher actions and tried to guess what they implied about student learning. Now we are going to turn this procedure around. Instead of asking what it means to teach mathematics, we are going to ask what it means to learn mathematics. Instead of looking at teacher actions and trying to read assumptions about students, (which is a risky matter, at best) we are going to look at some of the major assumptions about student differences and student learning patterns that have been proposed by psychologists and educators. We shall study mathematics teaching not by looking at teachers, but by looking at students. We cannot hope to understand teaching if we do not first understand the students who are to be taught.

Because the organization of this book is different from that of the usual textbook, you should take a few minutes to study its construction. The book is divided into major units. Each unit explores a major theme related to student abilities or the way students learn mathematics. Units are subdivided into learning modules. Each module centers around a basic activity related to the theme — an investigation, a reading, a discussion, an application of ideas to classroom practice, or a guide for further study.

In the investigation modules we shall do some simple experiments to see how people react to given learning situations. There is, after all, no substitute for first-hand experience. Original articles and writings in the reading modules show how psychologists and educators have used similar data and observations to formulate basic assumptions and theories about mathematics learning and learning abilities. Discussion modules relate accepted theories specifically to mathematics. Some kinds of teaching practices that one might expect to observe from a teacher who adopted these theories appear in application modules. Each unit concludes with a study module which suggests written activities, microteaching assignments, and additional reading resources. Study modules should be consulted frequently during the study of a unit.

Our modules differ from usual textbook chapters not only in their focus upon specific learning activities, but also in their open-ended nature. The approach is eclectic in order to reflect student diversity that teachers actually find in their mathematics classes. You should understand enough theories and practice enough teaching strategies to prepare for unexpected and varied teaching situations.

Unit 1

Intelligence and the Structure of Mathematical Abilities

Introduction

"I did quite well in school except for math; I just can't seem to learn mathematics!" How often have you heard a statement like that? Despite the fact that schools require every student to take at least six to eight years of course work in mathematics, it is still a common belief that some students have a mathematical ability and others do not. In this chapter we will look for an intelligence factor which might be called "mathematical ability." We will consider different ways in which that mental capacity called intelligence seems to be structured or patterned and the relations of these patterns to the ideas of mathematics.

Our work will be organized to emphasize learning by doing. Major ideas will be introduced by means of simple investigations. Unless otherwise noted, your classmates or friends can serve as the subjects for these experiments. You should not worry if the results of your experiments sometimes do not seem conclusive. Most of the experiments in this book are of current interest to researchers in psychological areas. They may not provide final answers, but they should raise interesting questions which will guide your further study. The study of human intelligence is not an area where anyone knows all the answers. Theories are still being born, tested, revised, accepted, and rejected. We will look at some of the most interesting ones which are currently vying for acceptance.

Is there a part of the brain which one uses to think mathematically and to learn mathematics? People thought so in the early 1800's. In studying anatomy and physiology, Franz Gall noticed that men with certain bumps or prominences on their heads possessed certain definite qualities. Gall claimed that mental qualities were associated with these physical characteristics, and the study of these associations came to be called *phrenology*. The temples on either side of the head supposedly contained the "organs" of number. Gall also decided that a certain bump gave poets their skills, and that other bumps made men thieves or murderers.

Today the theories of phrenology lay rejected and discarded. No serious researcher continues to look for an actual physical area of the brain which is related to a particular cognitive process. Physiology has shown that different portions of the brain do have special functions, but these are related to the translation of sensory stimuli to physical movements.

Then what makes one man a great mathematician and another a great poet? Although no one looks for the physical location of intellectual abilities, we are still trying to determine whether or not the intellect has naturally occurring parts or components. For example, if we could find great differences in individuals' abilities to learn, remember, and recall facts, and if the nature of these differences changed from one subject matter to another, we might consider intelligence to be a composite of several different abilities.

For example, suppose we gave people short paragraphs to read and then asked them a series of questions about the ideas contained in those paragraphs. If the paragraphs were equal in reading difficulty (that is, similar in vocabulary and in sentence construction), then differences in responses might reflect basic differences in the people's ability to understand ideas from different subject areas. Perhaps some would do well on paragraphs about history; others might do best on science paragraphs. Still others might do best on paragraphs containing mathematical ideas. We might begin to suspect that intelligence was actually a composite of many subject-matter abilities. An intelligence profile for a single individual might find him ranking high in the ability to learn English, low in the ability to learn history, high in the ability to learn science, average in the ability to learn mathematics, etc.

These test results would tend to confirm that some people have the ability to do mathematics while others simply do not. Constructing a test like the imaginary one would be a major undertaking. Not only would we have to equate the difficulty of the items, but we would have to see that each of them was an unfamiliar item so that differences were a result of differences in abilities and not in achievement in previous courses. However, we might be able to find supporting evidence for this basic idea by looking at simpler tests. In the following investigation module we will compare the ability of people to remember the basic elements of English with their ability to remember the basic elements of mathematics.

Mathematical Memory

Is it easier to remember a nonsense string of letters or a nonsense string of digits? Does the ability to remember one better than another vary from person to person? It should be relatively easy to find out. Suppose we read longer and longer sequences to an experimental subject and asked him to repeat them. We could record the longest sequence which was repeated correctly for letters and for digits. To improve the accuracy of the measurement, we will read an ascending series and a descending series, recording the longest sequence correctly repeated in each. The average of these two will be used as the measure of immediate memory span.

There are two cautions to be observed. There may be a difference between whether the digit span was read first or the letter span. For example, some subjects might do better on the second task, not because it was letters or digits, but because the first task had provided practice. To account for this practice give the digit span first with the first subject, the letter span first with the second subject, the digit span first with the third subject, etc. Then if we find the average digit memory span and the average letter memory span for several subjects, any systematic errors due to practice effect should cancel out.

In order to generalize the results of this experiment to a bigger group of people than just the subjects, some caution in selecting our subjects must be exercised. Subjects should be chosen at random from the group (population) to which we expect our results to apply. Suppose we want to know about the difference in immediate memory spans for letters and digits for seventh graders in the state of Illinois. How should we select our subjects? Obviously they should all be seventh graders, and all live in Illinois. But we must be more careful. How old are children in the seventh grade? The children we select should have a distribution of ages which parallels the actual distribution of ages for all seventh graders in Illinois. Similarly, if it is important that the conclusion be limited to Illinois students, then we should select a sample which reflects the geographical distribution of seventh graders over the state. In gross terms this means we should have more subjects from the Chicago area than from downstate Illinois. This selection process seems like a formidable task. In practice it is not. One solution would be to get a list of all seventh graders in Illinois and to choose the desired sample at random, perhaps by using a list of random numbers generated by a computer. Then, since most seventh graders are twelve years old, the probability that most of our sample would be twelve years old would be high. Since more seventh graders live in the Chicago area than in the rest of the state, there would be a higher probability of choosing subjects from the most populous area.

What does this example mean to the experiment at hand? In most cases you will not be able to choose subjects at random (in fact, the easiest way to do this experiment is simply to use your classmates as subjects). Therefore, extreme caution about generalizing the results must be exercised. If your subjects are all college seniors, to generalize about all the seniors at the college, depending on whether they were randomly selected or whether they selected themselves (for example, all students who elected to take the same course) may or may not be fair. Thus, some background data about each subject should be recorded before you actually perform the experiment. The exact nature of the data depends upon the general group of subjects selected. If you decide to use people in your dormitory, the following table might be useful.

Subject Number	Sex*	Age	Year Entered College	Major	Other Notes

*To be recorded if experiment is carried out in a coeducational dormitory.

Table 1. Background Data of Subjects.

Below are the digit-span and letter-span lists. Read each line slowly and clearly (about one symbol per second). Be sure you do not emphasize or connect any of the symbols by raising or lowering the tone of your voice. When you come to the end of the line raise your hand to signal the subject to respond.

Use the columns at the right to score the subject's response. Write a plus (+) if he is correct. Mark a (−) if he is wrong. Do not tell the subject if he is wrong, and do not stop until you have completed the entire list.

Digit Span	Subject Matter	1	2	3	4	5	6	7	8
972									
1406									
39418									
067285									
3516927									
58391204									
764580129									
2164089573									
45382170369									
870932614280									
541962836702									
39428307536									
4790386215									
042865129									
29081357									
1538796									
706294									
85943									
2730									
641									
Length of longest correct list in ascending series:									
Length of longest correct list in descending series:									
Average length (digit span):									

Table 2. Digit Span Record.

Letter Span	Subject Number	1	2	3	4	5	6	7	8
DVR									
FRXM									
PLBSN									
SDNWTC									
RTLMHTK									
GNKMBSCD									
LNSRPHTGB									
VLSMBCTPSH									
DRTSFHNRGKV									
CNRFDSRCTJXL									
BSVMLTPSCHRQ									
FTDMHKLCGBV									
RVDMXRFNSB									
LPCTWNDSK									
TMLBRDCM									
VRLDTNH									
XWPBFR									
MPVTS									
FDLH									
LMV									
Length of longest correct list in ascending series:									
Length of longest correct list in descending series:									
Average length (letter span):									

Table 3. Letter Span Record.

Analyzing the Data: Mean and Standard Deviation

You probably came to a decision about whether or not there was a difference in immediate memory span for digits and that for letters while you were still doing the experiment with your subjects. To see if the measurements recorded will tend to confirm or deny that decision, find the mean digit span and the mean letter span for all the subjects taken together. To find the (arithmetic) mean spans, total the average span lengths of all subjects and divide by the number of subjects.

Imagine that the value for the mean digit span is about the same as the value for the mean letter span. Would that mean the subjects had responded similarly with both digits and letters? Not necessarily. The subjects may have found the digits "easy" and remembered close to the mean number of digits; but at the same time they may have had "difficulty" with letters and given a wide variety of scores which just happened to average out to the same value. That is, while the mean is a measure of the central tendency of all the scores, it tells us nothing about how much the scores happen to be spread out.

One way to measure the "spread" of scores is to find the average of the difference between each score and the mean score for the group. However, this operation treats all these differences the same. Most people prefer to weight the average so that those scores which are farthest away from the mean influence the measure of spread the most. The simplest way to weight the average is to average the *square* of the difference between each score and the mean score. This average is called the *variance*. If the mean score is denoted by X, and there are n subjects in the group, then

$$\text{Variance} = \frac{\sum_{i=1}^{n}(\overline{X} - X_i)^2}{n}$$

Working with a measure of dispersion whose magnitude is more comparable to the magnitudes of the original scores is preferable. Thus, most researchers prefer to work with the square root of the value of the variance. This measure is called the *standard deviation* of the scores. That is

$$\text{Standard Deviation} = \sqrt{\text{Variance}} = \sqrt{\frac{\sum_{i=1}^{n}(\overline{X} - X_i)^2}{n}}$$

Compute the mean and standard deviation of the digit-span scores and the letter-span scores. The following table will help you organize your work.

Subject Number	Average Digit Span X	Deviation from Mean $d_x = \bar{X} - X$	Squared Deviation d_x^2		Average Letter Span Y	Deviation from Mean $d_y = \bar{Y} - Y$	Squared Deviation d_y^2
1							
2							
3							
4							
5							
6							
7							
8							

$$\sum X = \underline{\quad} \qquad\qquad \sum Y = \underline{\quad}$$

$$\bar{X} = \frac{\sum X}{n} = \underline{\quad} \qquad\qquad \bar{Y} = \frac{\sum Y}{n} = \underline{\quad}$$

$$\sum d_x^2 = \underline{\quad} \qquad\qquad \sum d_y^2 = \underline{\quad}$$

$$V_x = \frac{\sum d_x^2}{n} = \underline{\quad} \qquad\qquad V_y = \frac{\sum d_y^2}{n} = \underline{\quad}$$

$$SD_x = \sqrt{V_x} = \underline{\quad} \qquad\qquad SD_y = \sqrt{V_y} = \underline{\quad}$$

Table 4. Mean and Standard Deviation.

Analyzing the Data: Correlation

Suppose the means were about the same for both digit spans and letter spans, and the standard deviations were also approximately equal. Could we conclude that there is no difference in the ability to remember digits and the ability to remember letters? Not necessarily! If they are the same, then the ability to remember a long string of digits implies that the subject will also be able to remember a long string of letters, and vice versa. We ought to check for extremes — cases where the subject had a large digit span and a small letter span or vice versa. One way to check is to consider the digit and letter spans as an ordered pair for each individual and to plot them on a graph. Such a graph is called a scatter diagram. What pattern will the points form if digit and letter spans are strongly related in the same sense? What pattern will the points form if the relationship between the abilities is very weak? (In this situation the probability of pairing a high digit score with a low letter score is just as great as the probability of pairing high with high). Use the graph below to plot a scatter diagram for your data.

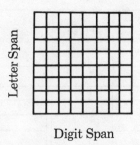

Digit Span

Figure 1. Scatter Diagram of Score.

A numerical measure of the relationship (or lack of relationship) shown on your scatter diagram is a *correlation coefficient*.
There are many ways to compute correlation coefficients. The formula below yields what is technically known as a product-moment-
correlation coefficient. We have chosen it because it employs the
same differences and standard deviations you tabulated and calculated on page 17. The formula for the correlation coefficient r is

$$r = \frac{\sum_{i=1}^{n} (d_{xi} \cdot d_{yi})}{n(SD_x)(SD_y)}$$

where the subscript x refers to digit spans and y refers to letter
spans. If the two abilities are highly correlated, the value of r will
approach positive one. If the two abilities are correlated in a negative sense (long digit spans are related to short letter spans), then
the value of r will approach negative one. If the relationship is
random, the value of r will approach zero.

Use the following table to organize your calculation of the correlation coefficient. Copy the data for the first two columns from
Table 4, being careful to include positive and negative signs. Use
the standard deviation values computed earlier.

Interpreting the Investigation

As you may have guessed as you worked through this investigation, psychologists have not been successful in finding naturally
occurring components of intelligence when they used subject matter as a basis for dividing and comparing. Most people who do this
experiment do not find great or persistent differences between the
ability of subjects to recall digits and their ability to recall letters.
After all, the two tasks are extremely similar. On the other hand,

Subject Number	Digit-Span Deviation from Mean d_x	Letter-Span Deviation from Mean d_y	Product $d_x \cdot d_y$
1			
2			
3			
4			
5			
6			
7			
8			

$$\sum_{i=1}^{n} (d_{xi} \cdot d_{yi}) = \underline{\hspace{3cm}}$$

$$r = \frac{\sum_{i=1}^{n} (d_{xi} \cdot d_{yi})}{n(SD_x)(SD_y)} = \underline{\hspace{3cm}}$$

Table 5. Calculation of Correlation Coefficient.

when differences do show up, they often have some interesting causes. You may have found some people who did very well at the digit task because they were able to add more information to the task. Perhaps they remembered the sequence by quickly relating successive digits to the preceding digits by thinking of addition or multiplication combinations. Or perhaps you found someone who did well with the letters because he inserted vowels in between to make up words or syllables. The use of schemes like this to aid in memory suggests that we find it easier to remember information if it has more meaning attached to it. Does this suggest implications for learning and teaching mathematics?

You may also have noticed a surprising consistency in the length of the span that subjects could recall. The average span is usually about seven digits or letters, which seems to represent our capacity to *immediately* store and retrieve information. If we are to store and retrieve longer spans of information, we must interject some kind of memory aid by consolidating the information, by relating it to something else we already know, or by simply repeating it over and over until it is easier to recall.

Do these findings mean that intelligence cannot be subdivided? There are other ways to divide and compare tasks. Suppose we compared the ability to recall a list of symbols (either letters or digits) with the ability to recall a sequence of the orientations of a given object in space. Would this division of content show differences in ability?

Suppose that we compared the ability to recall a list of digits read aloud with the ability to recall a similar list projected on a screen. Would we be able to discover a difference in recall ability according to the method of presentation?

You may want to try experiments like these for yourself. Trying such experiments in a very informal way is one of the most exciting aspects of classroom teaching. Formal experiments, similar to these, have occupied one branch of psychology for much of this century. The following reading module contains an article by the leading researcher in the field of intelligence testing, J. P. Guilford. In the article, Professor Guilford summarizes and organizes much of what we know about human intellectual abilities today. In addition, he presents a model of the intellect which predicts the existence of several abilities whose measures we have still not discovered.

Three Faces
of Intellect

J.P.Guilford

My subject is in the area of human intelligence, in connection with which the names of Terman and Stanford have become known the world over. The Stanford Revision of the Binet intelligence scale has been the standard against which all other instruments for the measurement of intelligence have been compared. The term IQ or intelligence quotient has become a household word in this country. This is illustrated by two brief stories.

A few years ago, one of my neighbors came home from a PTA meeting, remarking: "That Mrs. So-And-So, thinks she knows so much. She kept talking about the 'intelligence *quota*' of the children; 'intelligence *quota*'; imagine. Why, everybody knows that IQ stands for "intelligence *quiz*.'"

The other story comes from a little comic strip in a Los Angeles morning newspaper, called "Junior Grade." In the first picture a little boy meets a little girl, both apparently about the first-grade level. The little girl remarks, "I have a high I.Q." The little boy, puzzled, said, "You have a what?" The little girl repeated, "I have a high IQ," then went on her way. The little boy, looking thoughtful, said, "And she looks like such a nice little girl, too."

It is my purpose to speak about the analysis of this thing called human intelligence into its components. I do not believe that either Binet or Terman, if they were still with us, would object to the idea of a search and detailed study of intelligence, aimed toward a better understanding of its nature. Preceding the development of his intelligence scale, Binet had done much research on different kinds of thinking activities and apparently recognized that intelligence has a number of aspects. It is to the lasting credit of both Binet and Terman that they introduced such a variety of tasks into their intelligence scales.

Two related events of very recent history make it imperative that we learn all we can regarding the nature of intelligence. I am referring to the advent of the artificial satellites and planets and to the crisis in education that has arisen in part as a consequence. The preservation of our way of life and our future security depend upon our most important national resources: our intellectual abilities and, more particularly, our creative abilities. It is time, then, that we learn all we can about those resources.

Our knowledge of the components of human intelligence has come about mostly within the last 25 years. The major sources of this information in this country have been L. L. Thurstone and his associates, the wartime research of psychologists in the United States Air Forces, and

Reprinted from *American Psychologist,* 1959, 14, 469-479, by permission of the author and the American Psychological Association; J. P. Guilford, *Three Faces of Intellect.*

more recently the Aptitudes Project[1] at the University of Southern California, now in its tenth year of research on cognitive and thinking abilities. The results from the Aptitudes Project that have gained perhaps the most attention have pertained to creative-thinking abilities. These are mostly novel findings. But to me, the most significant outcome has been the development of a unified theory of human intellect, which organizes the known, unique, or primary intellectual abilities into a single system called the "structure of intellect." It is to this system that I shall devote the major part of my remarks, with very brief mentions of some of the implications for the psychology of thinking and problem solving, for vocational testing, and for education.

The discovery of the components of intelligence has been by means of the experimental application of the method of factor analysis. It is not necessary for you to know anything about the theory or method of factor analysis in order to follow the discussion of the components. I should like to say, however, that factor analysis has no connection with or resemblance to psychoanalysis. A positive statement would be more helpful, so I will say that each intellectual component or factor is a unique ability that is needed to do well in a certain class of tasks or tests. As a general principle we find that certain individuals do well in the tests of a certain class, but they may do poorly in the tests of another class. We conclude that a factor has certain properties from the features that the tests of a class have in common. I shall give you very soon a number of examples of tests, each representing a factor.

The Structure of Intellect

Although each factor is sufficiently distinct to be detected by factor analysis, in very recent years it has become apparent that the factors themselves can be classified because they resemble one another in certain ways. One basis of classification is according to the basic kind of process or operation performed. This kind of classification gives us five major groups of intellectual abilities: factors of cognition, memory, convergent thinking, divergent thinking, and evaluation.

Cognition means discovery or rediscovery or recognition. Memory means retention of what is cognized. Two kinds of productive-thinking operations generate new information from known information and remembered information. In divergent-thinking operations we think in different directions, sometimes searching, sometimes seeking variety. In convergent thinking the information leads to one right answer or to a recognized best or conventional answer. In evaluation we reach decisions as to goodness, correctness, suitability, or adequacy of what we know, what we remember, and what we produce in productive thinking.

A second way of classifying the intellectual factors is according to the kind of material or content involved. The factors known thus far involve three kinds of material or content: the content may be figural, symbolic, or semantic. Figural content is concrete material such as is perceived through the senses. It does not represent anything except itself. Visual material has properties such as size, form, color, location, or texture.

---

[1]Under Contract N6onr-23810 with the Office of Naval Research (Personnel and Training Branch).

Things we hear or feel provide other examples of figural material. Symbolic content is composed of letters, digits, and other conventional signs, usually organized in general systems, such as the alphabet or the number system. Semantic content is in the form of verbal meanings or ideas, for which no examples are necessary.

When a certain operation is applied to a certain kind of content, as many as six general kinds of products may be involved. There is enough evidence available to suggest that, regardless of the combinations of operations and content, the same six kinds of products may be found associated. The six kinds of products are: units, classes, relations, systems, transformations, and implications. So far as we have determined from factor analysis, these are the only fundamental kinds of products that we can know. As such, they may serve as basic classes into which one might fit all kinds of information psychologically.

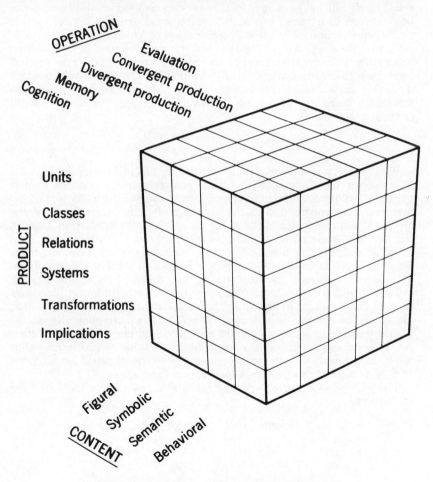

Figure 2. The Structure of Intellect.

A cubical model representing the structure of intellect.

The three kinds of classifications of the factors of intellect can be represented by means of a single solid model, shown in Figure 2. In this model, which we call the "structure of intellect," each dimension represents one of the modes of variation of the factors.[2] Along one dimension are found the various kinds of operations, along a second one are the various kinds of products, and along the third are various kinds of content. Along the dimension of content a fourth category has been added, its kind of content being designated as "behavioral." This category has been added on a purely theoretical basis to represent the general area sometimes called "social intelligence." More will be said about this section of the model later.

In order to provide a better basis for understanding the model and a better basis for accepting it as a picture of human intellect, I shall do some exploring of it with you systematically, giving some examples of tests. Each cell in the model calls for a certain kind of ability that can be described in terms of operation, content, and product, for each cell is at the intersection of a unique combination of kinds of operation, content, and product. A test for that ability would have the same three properties. In our exploration of the model, we shall take one vertical layer at a time, beginning with the front face. The first layer provides us with a matrix of eighteen cells (if we ignore the behavioral column for which there are as yet no known factors) each of which should contain a cognitive ability.

The Cognitive Abilities

We know at present the unique abilities that fit logically into 15 of the 18 cells for cognitive abilities. Each row presents a triad of similar abilities, having a single kind of product in common. The factors of the first row are concerned with the knowing of units. A good test of the ability to cognize figural units is the Street Gestalt Completion Test. In this test, the recognition of familiar pictured objects in silhouette form is made difficult for testing purposes by blocking out parts of those objects. There is another factor that is known to involve the perception of auditory figures — in the form of melodies, rhythms, and speech sounds — and still another factor involving kinesthetic forms. The presence of three factors in one cell (they are conceivably distinct abilities, although this has not been tested) suggests that more generally, in the figural column, at least, we should expect to find more than one ability. A fourth dimension pertaining to variations in sense modality may thus apply in connection with figural content. The model could be extended in this manner if the facts call for such an extension.

The ability to cognize symbolic units is measured by tests like the following:

Put vowels in the following blanks to make real words:

    P__W__R
    M__RV__L
    C__TR__N

---

[2]For an earlier presentation of the concept, see Guilford (1956).

Rearrange the letters to make real words:

    R A C I H
    T V O E S
    K L C C O

The first of these two tests is called Disemvoweled Words, and the second Scrambled Words.

The ability to cognize semantic units is the well-known factor of verbal comprehension, which is best measured by means of a vocabulary test, with items such as:

GRAVITY means _____

CIRCUS means _____

VIRTUE means _____

From the comparison of these two factors it is obvious that recognizing familiar words as letter structures and knowing what words mean depend upon quite different abilities.

For testing the abilities to know classes of units, we may present the following kinds of items, one with symbolic content and one with semantic content:

Which letter group does not belong?

    XECM   PVAA   QXIN   VTRO

Which object does not belong?

    clam   tree   oven   rose

A figural test is constructed in a completely parallel form, presenting in each item four figures, three of which have a property in common and the fourth lacking that property.

The three abilities to see relationship are also readily measured by a common kind of test, differing only in terms of content. The well-known analogies test is applicable, two items in symbolic and semantic form being:

    JIRE   : KIRE   : : FORA : KORE   KORA LIRE GORA GIRE
    poetry : prose : : dance : music   walk sing talk jump

Such tests usually involve more than the ability to cognize relations, but we are not concerned with this problem at this point.

The three factors for cognizing systems do not at present appear in tests so closely resembling one another as in the case of the examples just given. There is nevertheless an underlying common core of logical similarity. Ordinary space tests, such as Thurstone's Flags, Figures, and Cards or Part V (Spatial Orientation) of the Guilford-Zimmerman Aptitude Survey (GZAS), serve in the figural column. The system involved is an order or arrangement of objects in space. A system that uses symbolic elements is illustrated by the Letter Triangle Test, a sample of which is:

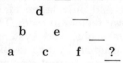

What letter belongs at the place of the question mark?

The ability to understand a semantic system has been known for some time as the factor called general reasoning. One of its most faithful indicators is a test composed of arithmetic-reasoning items. That the phase of understanding only is important for measuring this ability is shown by the fact that such a test works even if the examinee is not asked to give a complete solution; he need only show that he structures the problem properly. For example, an item from the test Necessary Arithmetical Operations simply asks what operations are needed to solve the problem:

A city lot 48 feet wide and 149 feet deep costs $79,432. What is the cost per square foot?	A. add and multiply B. multiply and divide C. subtract and divide D. add and subtract E. divide and add

Placing the factor of general reasoning in this cell of the structure of intellect gives us some new conceptions of its nature. It should be a broad ability to grasp all kinds of systems that are conceived in terms of verbal concepts, not restricted to the understanding of problems of an arithmetical type.

Transformations are changes of various kinds, including modifications in arrangement, organization, or meaning. In the figural column for the transformations row, we find the factor known as visualization. Common measuring instruments for this factor are the surface-development tests, and an example of a different kind is Part VI (Spatial Visualization) of the GZAS. A test of the ability to make transformations of meaning, for the factor in the semantic column, is called Similarities. The examinee is asked to state several ways in which two objects, such as an apple and an orange, are alike. Only by shifting the meanings of both is the examinee able to give many responses to such an item.

In the set of abilities having to do with the cognition of implications, we find that the individual goes beyond the information given, but not to the extent of what might be called drawing conclusions. We may say that he extrapolates. From the given information he expects or foresees certain consequences, for example. The two factors found in this row of the cognition matrix were first called "foresight" factors. Foresight in connection with figural material can be tested by means of paper-and-pencil mazes. Foresight in connection with ideas, those pertaining to events, for example, is indicated by a test such as Pertinent Questions:

In planning to open a new hamburger stand in a certain community, what four questions should be considered in deciding upon its location? The more questions the examinee asks in response to a list of such problems, the more he evidently foresees contingencies.

The Memory Abilities

The area of memory abilities has been explored less than some of the other areas of operation, and only seven of the potential cells of the memory matrix have known factors in them. These cells are restricted to three rows: for units, relations, and systems. The first cell in the memory matrix is now occupied by two factors, parallel to two in the corresponding cognition matrix: visual memory and auditory memory. Memory for series of letters or numbers, as in memory-span tests, conforms to the conception of memory for symbolic units. Memory for the ideas in a paragraph conforms to the conception of memory for semantic units.

The formation of associations between units, such as visual forms, syllables, and meaningful words, as in the method of paired associates, would seem to represent three abilities to remember relationships involving three kinds of content. We know of two such abilities, for the symbolic and semantic columns. The memory for known systems is represented by two abilities very recently discovered (Christal, 1958). Remembering the arrangement of objects in space is the nature of an ability in the figural column, and remembering a sequence of events is the nature of a corresponding ability in the semantic column. The differentiation between these two abilities implies that a person may be able to say where he saw an object on a page, but he might not be able to say on which of several pages he saw it after leafing through several pages that included the right one. Considering the blank rows in the memory matrix, we should expect to find abilities also to remember classes, transformations, and implications, as well as units, relations, and systems.

The Divergent-Thinking Abilities

The unique feature of divergent production is that a *variety* of responses is produced. The product is not completely determined by the given information. This is not to say that divergent thinking does not come into play in the total process of reaching a unique conclusion, for it comes into play wherever there is trial-and-error thinking.

The well-known ability of word fluency is tested by asking the examinee to list words satisfying a specified letter requirement, such as words beginning with the letter "s" or words ending in "-tion." This ability is now regarded as a facility in divergent production of symbolic units. The parallel semantic ability has been known as ideational fluency. A typical test item calls for listing objects that are round and edible. Winston Churchill must have possessed this ability to a high degree. Clement Atlee is reported to have said about him recently that, no matter what problem came up, Churchill always seemed to have about ten ideas. The trouble was, Attlee continued, he did not know which was the good one. The last comment implies some weakness in one or more of the evaluative abilities.

The divergent production of class ideas is believed to be the unique feature of a factor called "spontaneous flexibility." A typical test instructs the examinee to list all the uses he can think of for a common brick, and he is given eight minutes. If his responses are: build a house, build a barn, build a garage, build a school, build a church, build a chimney, build a walk, and build a barbecue, he would earn a fairly high score for ideational fluency but a very low score for spontaneous flexibility, because all these uses fall into the same class. If another person said: make a door stop, make a paper weight, throw it at a dog, make a bookcase, drown a cat, drive a nail, make a red powder, and use for baseball bases, he would also receive a high score for flexibility. He has gone frequently from one class to another.

A current study of unknown but predicted divergent-production abilities includes testing whether there are also figural and symbolic abilities to produce multiple classes. An experimental figural test presents a number of figures that can be classified in groups of three in various ways, each figure being usable in more than one class. An experimental symbolic test presents a few numbers that are also to be classified in multiple ways.

A unique ability involving relations is called "associational fluency." It calls for the production of a variety of things related in a specified way to a given thing. For example, the examinee is asked to list words meaning about the same as "good" or to list words meaning about the opposite of "hard." In these instances the response produced is to complete a relationship, and semantic content is involved. Some of our present experimental tests call for the production of varieties of relations, as such, and involve figural and symbolic content also. For example, given four small digits, in how many ways can they be related in order to produce a sum of eight?

One factor pertaining to the production of systems is known as expressional fluency. The rapid formation of phrases or sentences is the essence of certain tests of this factor. For example, given the initial letters:

W_____ c_____ e_____ n_____

with different sentences to be produced, the examinee might write "We can eat nuts" or "Whence came Eve Newton?" In interpreting the factor, we regard the sentence as a symbolic system. By analogy, a figural system would be some kind of organization of lines and other elements, and a semantic system would be in the form of a verbally stated problem or perhaps something as complex as a theory.

In the row of the divergent-production matrix devoted to transformations, we find some very interesting factors. The one called "adaptive flexibility" is now recognized as belonging to the figural column. A faithful test of it has been Match Problems. This is based upon the common game that uses squares, the sides of which are formed by match sticks. The examinee is told to take away a given number of matches to leave a stated number of squares with nothing left over. Nothing is said about the sizes of the squares to be left. If the examinee imposes

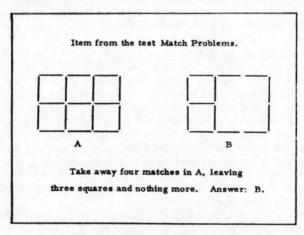

Figure 3. Match Problems Test.

A sample item from the test Match Problems. The problem in this item is to take away four matches and leave three squares. The solution is given.

upon himself the restriction that the squares that he leaves must be of the same size, he will fail in his attempts to do items like that in Figure 3. Other odd kinds of solutions are introduced in other items, such as overlapping squares and squares within squares, and so on. In another variation of Match Problems the examinee is told to produce two or more solutions for each problem.

A factor that has been called "originality" is now recognized as adaptive flexibility with semantic material, where there must be a shifting of meanings. The examinee must produce the shifts or changes in meaning and so come up with novel, unusual, clever, or farfetched ideas. The Plot Titles Test presents a short story, the examinee being told to list as many appropriate titles as he can to head the story. One story is about a missionary who has been captured by cannibals in Africa. He is in the pot and about to be boiled when a princess of the tribe obtains a promise for his release if he will become her mate. He refuses and is boiled to death.

In scoring the test, we separate the responses into two categories, clever and nonclever. Examples of nonclever responses are: African Death, Defeat of a Princess, Eaten by Savages, The Princess, The African Missionary, In Darkest Africa, and Boiled by Savages. These titles are appropriate but commonplace. The number of such responses serves as a score for ideational fluency. Examples of clever responses are: Pot's Plott, Potluck Dinner, Stewed Parson, Goil or Boil, A Mate Worse Than Death, He Left a Dish for a Pot, Chaste in Haste, and A Hot Price for Freedom. The number of clever responses given by an examinee is his score for originality, or the divergent production of semantic transformations.

Another test of originality presents a very novel task so that any acceptable response is unusual for the individual. In the Symbol Production Test the examinee is to produce a simple symbol to stand for a noun or a verb in each short sentence, in other words to invent something like pictograph symbols. Still another test of originality asks for writing the "punch lines" for cartoons, a task that almost automatically challenges the examinee to be clever. Thus, quite a variety of tests offer approaches to the measurement of originality, including one or two others that I have not mentioned.

Abilities to produce a variety of implications are assessed by tests calling for elaboration of given information. A figural test of this type provides the examinee with a line or two, to which he is to add other lines to produce an object. The more lines he adds, the greater his score. A semantic test gives the examinee the outlines of a plan to which he is to respond by stating all the details he can think of to make the plan work. A new test we are trying out in the symbolic area presents two simple equations such as $B - C = D$ and $z = A + D$. The examinee is to make as many other equations as he can from this information.

The Convergent-Production Abilities

Of the 18 convergent-production abilities expected in the three content columns, 12 are now recognized. In the first row, pertaining to units, we have an ability to name figural properties (form or colors) and an ability to name abstractions (classes, relations, and so on). It may be that the ability in common to the speed of naming forms and the speed of naming colors is not appropriately placed in the convergent-thinking matrix. One might expect that the thing to be produced in a test of the convergent production of figural units would be in the form of figures rather than words. A better test of such an ability might somehow specify the need for one particular object, the examinee to furnish the object.

A test for the convergent production of classes (Word Grouping) presents a list of 12 words that are to be classified in four, and only four, meaningful groups, no word to appear in more than one group. A parallel test (Figure Concepts Test) presents 20 pictured real objects that are to be grouped in meaningful classes of two or more each.

Convergent production having to do with relationships is represented by three known factors, all involving the "education of correlates," as Spearman called it. The given information includes one unit and a stated relation, the examinee to supply the other unit. Analogies tests that call for completion rather than a choice between alternative answers emphasize this kind of ability. With symbolic content such an item might read:

pots   stop   bard   drab   rats   ?

A semantic item that measures education of correlates is:

The absence of sound is _____.

Incidentally, the latter item is from a vocabulary-completion test, and its relation to the factor of ability to produce correlates indicates how, by change of form, a vocabulary test may indicate an ability other than that for which vocabulary tests are usually intended, namely, the factor of verbal comprehension.

Only one factor for convergent production of systems is known, and it is in the semantic column. It is measured by a class of tests that may be called ordering tests. The examinee may be presented with a number of events that ordinarily have a best or most logical order, the events being presented in scrambled order. The presentation may be pictorial, as in the Picture Arrangement Test, or verbal. The pictures may be taken from a cartoon strip. The verbally presented events may be in the form of the various steps needed to plant a new lawn. There are undoubtedly other kinds of systems than temporal order that could be utilized for testing abilities in this row of the convergent-production matrix.

In the way of producing transformations of a unique variety, we have three recognized factors, known as redefinition abilities. In each case, redefinition involves the changing of functions or uses of parts of one unit and giving them new functions or uses in some new unit. For testing the ability of figural redefinition, a task based upon the Gottschaldt figures is suitable. Figure 4 shows the kind of item for such a test. In recognizing the simpler figure within the structure of a more complex figure, certain lines must take on new roles.

In terms of symbolic material, the following sample items will illustrate how groups of letters in given words must be readapted to use in other words. In the test Camouflaged Words, each sentence contains the name of a sport or game:

I did not know that he was ailing.

To beat the Hun, tin goes a long way.

For the factor of semantic redefinition, the Gestalt Transformation Test may be used. A sample item reads:
From which object could you most likely make a needle?
    A. a cabbage
    B. a splice
    C. a steak
    D. a paper box
    E. a fish

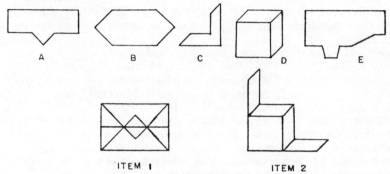

Figure 4. Hidden Figures Test.

Sample items from a test Hidden Figures, based upon the Gottschaldt figures. Which of the simpler figures is concealed within each of the two more complex figures?

The convergent production of implications means the drawing of fully determined conclusions from given information. The well-known factor of numerical facility belongs in the symbolic column. For the parallel ability in the figural column, we have a test known as Form Reasoning, in which rigorously defined operations with figures are used. For the parallel ability in the semantic column, the factor sometimes called "deduction" probably qualifies. Items of the following type are sometimes used.

        Charles is younger than Robert
        Charles is older than Frank
        Who is older: Robert or Frank?

Evaluative Abilities

The evaluative area has had the least investigation of all the operational categories. In fact, only one systematic analytical study has been devoted to this area. Only eight evaluative abilities are recognized as fitting into the evaluation matrix. But at least five rows have one or more factors each, and also three of the usual columns or content categories. In each case, evaluation involves reaching decisions as to the accuracy, goodness, suitability, or workability of information. In each row, for the particular kind of product of that row, some kind of criterion or standard of judgment is involved.

In the first row, for the evaluation of units, the important decision to be made pertains to the identity of a unit. Is this unit identical with that one? In the figural column we find the factor long known as "perceptual speed." Tests of this factor invariably call for decisions of identity, for example, Part IV (Perceptual Speed) of the GZAS or Thurstone's Identical Forms. I think it has been generally wrongly thought that the ability involved is that of cognition of visual forms. But we have seen that another factor is a more suitable candidate for this definition and for being in the very first cell of the cognitive matrix. It is parallel to this evaluative ability but does not require the judgment of identity as one of its properties.

In the symbolic column is an ability to judge identity of symbolic units, in the form of series of letters or numbers or of names of individuals.

Are members of the following pairs identical or not:

825170493_____825176493
dkeltvmpa_____dkeltvmpa
C. S. Myerson_____C. E. Myerson

Such items are common in tests of clerical aptitude.

There should be a parallel ability to decide whether two ideas are identical or different. Is the idea expressed in this sentence the same as the idea expressed in that one? Do these two proverbs express essentially the same idea? Such tests exist and will be used to test the hypothesis that such an ability can be demonstrated.

No evaluative abilities pertaining to classes have as yet been recognized. The abilities having to do with evaluation where relations are concerned must meet the criterion of logical consistency. Syllogistic-type tests involving letter symbols indicate a different ability than the same type of test involving verbal statements. In the figural column we might expect that tests incorporating geometric reasoning or proof would indicate a parallel ability to sense the soundness of conclusions regarding figural relationships.

The evaluation of systems seems to be concerned with the internal consistency of those systems, so far as we can tell from the knowledge of one such factor. The factor has been called "experiential evaluation," and its representative test presents items like that in Figure 5 asking "What is wrong with this picture?" The things wrong are often internal inconsistencies.

A semantic ability for evaluating transformations is thought to be that known for some time as "judgment." In typical judgment tests, the examinee is asked to tell which of five solutions to a practical problem is most adequate or wise. The solutions frequently involve improvisations, in other words, adaptations of familiar objects to unusual uses. In this way the items present redefinitions to be evaluated.

A factor known first as "sensitivity to problems" has become recognized as an evaluative ability having to do with implications. One test of the factor, the Apparatus Test, asks for two needed improvements with respect to each of several common devices, such as the telephone or the toaster. The Social Institutions Test, a measure of the same factor, asks what things are wrong with each of several institutions, such as tipping or national elections. We may say that defects or deficiencies are implications of an evaluative kind. Another interpretation would be that seeing defects and deficiencies are evaluations of implications to the effect that the various aspects of something are all right.[3]

---

[3]For further details concerning the intellectual factors, illustrative tests, and the place of the factors in the structure of intellect; see Guilford (1959).

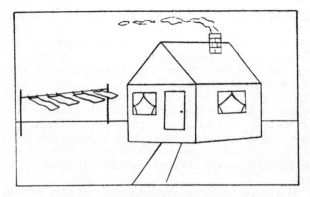

Figure 5. Unusual Details Test.

*A sample item from the test Unusual Details. What two
things are wrong with this picture?*

Some Implications of the Structure of Intellect

For Psychological Theory

Although factor analysis as generally employed is best designed to
investigate ways in which individuals differ from one another, in other
words, to discover traits, the results also tell us much about how individuals are alike. Consequently, information regarding the factors and
their interrelationships gives us understanding of functioning individuals. The five kinds of intellectual abilities in terms of operations may
be said to represent five ways of functioning. The kinds of intellectual
abilities distinguished according to varieties of test content and the
kinds of abilities distinguished according to varieties of products suggest
a classification of basic forms of information or knowledge. The kind
of organism suggested by this way of looking at intellect is that of an
agency for dealing with information of various kinds in various ways.
The concepts provided by the distinctions among the intellectual abilities and by their classifications may be very useful in our future investigations of learning, memory, problem solving, invention, and decision
making, by whatever method we choose to approach those problems.

For Vocational Testing

With about 50 intellectual factors already known, we may say that
there are at least 50 ways of being intelligent. It has been facetiously
suggested that there seem to be a great many more ways of being
stupid, unfortunately. The structure of intellect is a theoretical model
that predicts as many as 120 distinct abilities, if every cell of the model
contains a factor. Already we know that two cells contain two or more
factors each, and there probably are actually other cells of this type.
Since the model was first conceived, 12 factors predicted by it have
found places in it. There is consequently hope of filling many of the
other vacancies, and we may eventually end up with more than 120
abilities.

The major implication for the assessment of intelligence is that to
know an individual's intellectual resources thoroughly we shall need a

surprisingly large number of scores. It is expected that many of the factors are intercorrelated, so there is some possibility that by appropriate sampling we shall be able to cover the important abilities with a more limited number of tests. At any rate, a multiple-score approach to the assessment of intelligence is definitely indicated in connection with future vocational operations.

Considering the kinds of abilities classified as to content, we may speak roughly of four kinds of intelligence. The abilities involving the use of figural information may be regarded as "concrete" intelligence. The people who depend most upon these abilities deal with concrete things and their properties. Among these people are mechanics, operators of machines, engineers (in some aspects of their work), artists, and musicians.

In the abilities pertaining to symbolic and semantic content, we have two kinds of "abstract" intelligence. Symbolic abilities should be important in learning to recognize words, to spell, and to operate with numbers. Language and mathematics should depend very much upon them, except that in mathematics some aspects, such as geometry, have strong figural involvement. Semantic intelligence is important for understanding things in terms of verbal concepts and hence is important in all courses where the learning of facts and ideas is essential.

In the hypothesized behavioral column of the structure of intellect, which may be roughly described as "social" intelligence, we have some of the most interesting possibilities. Understanding the behavior of others and of ourselves is largely nonverbal in character. The theory suggests as many as 30 abilities in this area, some having to do with understanding, some with productive thinking about behavior, and some with the evaluation of behavior. The theory also suggests that information regarding behavior is also in the form of the six kinds of products that apply elsewhere in the structure of intellect, including units, relations, systems, and so on. The abilities in the area of social intelligence, whatever they prove to be, will possess considerable importance in connection with all those individuals who deal most with other people: teachers, law officials, social workers, therapists, politicians, statesmen, and leaders of other kinds.

For Education

The implications for education are numerous, and I have time just to mention a very few. The most fundamental implication is that we might well undergo transformations with respect to our conception of the learner and of the process of learning. Under the prevailing conception, the learner is a kind of stimulus-response device, much on the order of a vending machine. You put in a coin, and something comes out. The machine learns what reaction to put out when a certain coin is put in. If, instead, we think of the learner as an agent for dealing with information, where information is defined very broadly, we have something more analogous to an electronic computor. We feed a computor information; it stores that information; it uses that information for generating new information, either by way of divergent or convergent thinking; and it evaluates its own results. Advantages that a human learner has over a computor include the step of seeking and discovering new information from sources outside itself and the step of programming itself. Perhaps even these steps will be added to computors, if this has not already been done in some cases.

At any rate, this conception of the learner leads us to the idea that learning is discovery of information, not merely the formation of associations, particularly associations in the form of stimulus-response connections. I am aware of the fact that my proposal is rank heresy. But if we are to make significant progress in our understanding of human learning and particularly our understanding of the so-called higher mental processes of thinking, problem solving, and creative thinking, some drastic modifications are due in our theory.

The idea that education is a matter of training the mind or of training the intellect has been rather unpopular, wherever the prevailing psychological doctrines have been followed. In theory, at least, the emphasis has been upon the learning of rather specific habits or skills. If we take our cue from factor theory, however, we recognize that most learning probably has both specific and general aspects or components. The general aspects may be along the lines of the factors of intellect. This is not to say that the individual's status in each factor is entirely determined by learning. We do not know to what extent each factor is determined by heredity and to what extent by learning. The best position for educators to take is that possibly every intellectual factor can be developed in individuals at least to some extent by learning.

If education has the general objective of developing the intellects of students, it can be suggested that each intellectual factor provides a particular goal at which to aim. Defined by a certain combination of content, operation, and product, each goal ability then calls for certain kinds of practice in order to achieve improvement in it. This implies choice of curriculum and the choice or invention of teaching methods that will most likely accomplish the desired results.

Considering the very great variety of abilities revealed by the factorial exploration of intellect, we are in a better position to ask whether any general intellectual skills are now being neglected in education and whether appropriate balances are being observed. It is often observed these days that we have fallen down in the way of producing resourceful, creative graduates. How true this is, in comparison with other times, I do not know. Perhaps the deficit is noticed because the demands for inventiveness are so much greater at this time. At any rate, realization that the more conspicuously creative abilities appear to be concentrated in the divergent-thinking category, and also to some extent in the transformation category, we now ask whether we have been giving these skills appropriate exercise. It is probable that we need a better balance of training in the divergent-thinking area as compared with training in convergent thinking and in critical thinking or evaluation.

The structure of intellect as I have presented it to you may or may not stand the test of time. Even if the general form persists, there are likely to be some modifications. Possibly some different kind of model will be invented. Be that as it may, the fact of a multiplicity of intellectual abilities seems well established.

There are many individuals who long for the good old days of simplicity, when we got along with one unanalyzed intelligence. Simplicity certainly has its appeal. But human nature is exceedingly complex, and we may as well face that fact. The rapidly moving events of the world in which we live have forced upon us the need for knowing human intelligence thoroughly. Humanity's peaceful pursuit of happiness depends upon our control of nature and of our own behavior; and this, in turn, depends upon understanding ourselves, including our intellectual resources.

References

Christal, R. E. Factor analytic study of visual memory. *Psychol. Monogr.*, 1958, 72, No. 13 (Whole No. 466).

Guilford, J. P. The structure of intellect. *Psychol. Bull.*, 1956, 53, 267-293.

Guilford, J. P. *Personality*. New York: McGraw-Hill, 1959.

The Structure
of the Intellect
and the Structure
of Mathematical
Abilities

Guilford finds as many as 120 different factors which make up human intelligence. How are these factors reflected in mathematical ability? If intelligence is multifaceted, the ability in mathematics is certainly no less multifaceted. We can translate almost all the categories of each of Guilford's three dimensions into corresponding categories for mathematics ability. We shall discuss the implications of each of the three dimensions for mathematics in this module.

Content Abilities in Mathematics

Consider the *Contents* dimension of Guilford's model. Guilford identifies four categories by which ideas are communicated and received. These are figural, symbolic, semantic, and behavioral. We can find instances of three of these categories in mathematics. The behavioral category, or social behavior, is not of major importance in mathematics. Mathematics is a subject of ideas and relationships between ideas. It is abstract in the sense that the ideas of mathematics do not *necessarily* refer to physical objects or to symbolic conventions. In this sense, mathematics is *semantic*. "Semantic information is in the form of meanings to which words commonly become attached; hence, it is the most notable in verbal thinking and verbal communications." (2:227)* It is not accidental, nor a perversion by our educational system, that much of mathematics learning ultimately involves the learning of vocabulary. Words provide the best way we have to communicate mathematical ideas once we have attached the proper level of meaning to these words. The process of making mathematical terms meaningful is a major part of mathematics teaching. Thus, differences in the ability to process information received in the semantic mode creates differences in ability to learn and communicate mathematical ideas. As modern

---

*Numbers within parentheses refer to References at the end of each unit.

mathematics programs place more and more stress on ideas (as opposed to manipulative skills), semantic abilities must play a greater role in mathematics ability.

But mathematics is not comprised of words alone. One of the most important features of mathematics is its power as a symbolic system. We can transform ideas by representing them as symbols, manipulating the symbols according to logically based rules and conventions, and translating from the symbolic result to the new idea. This is a powerful technique, indeed. When we find that $(x+3)(x+5) = x(x+5) + 3(x+5) = x^2 + 5x + 3x + 15 = x^2 + 8x + 15$, we are handling by means of symbols a transformation of ideas about compound quantities under multiplication which would be very hard to learn and understand at a purely semantic level. Mathematics books from the sixteenth and seventeenth centuries are very difficult to understand because symbolic systems had not been well developed at that stage. The ability to translate ideas to symbols and symbols back to ideas is a most important facet of mathematical ability.

Symbols are signs which have no inherent meaning in themselves. Thus, nothing in the symbol "2" itself implies the idea of *twoness*. Indeed, the symbol is not even symmetric. Similarly, there is no idea of *fiveness* in the symbol "5." Nevertheless, some young children do build up a system by which number is imbedded within numerals. The system usually develops something like this: the child begins to focus on the center of the numeral "1" until he gradually sees it with one dot attached, 1. In a similar fashion he focuses on the top and bottom of the numeral "2" so it becomes 2. The addition problem $\begin{smallmatrix}2\\+1\end{smallmatrix}$ becomes the problem $\begin{smallmatrix}2\\+1\end{smallmatrix}$ which is solved by counting the dots to yield the result 3. Dots are attached to other numerals something like the following scheme.

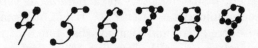

Figure 6. Learning Numerals as Symbols.

The system is cumbersome beyond six; it requires a more complex algorithm for numbers beyond nine. What the child has done, however, is to transform the symbols, which have no meaning in their own forms, into figures which do represent information in a concrete or physical form. This process is the last important way in which mathematics is learned and transmitted: in a figural mode.

Like words and symbols, figures, pictures, and graphs play an important role in mathematics. The power of figures resides in the power of spatial representation. Topology and geometry are, of course, primarily concerned with spatial ideas. But spatial representation is a useful tool even in analyzing nonspatial ideas. Thus, the concept of set inclusion (semantic), represented by $A \subset B$ (symbolic), is transmitted perhaps best by the Venn diagram

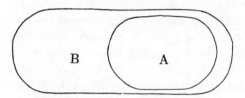

Figure 7. Concept of Set Inclusion.

Differing abilities to think, interpret, and organize ideas in figural ways undoubtedly leads to different ability levels in mathematics.

The implications for the classroom teacher of mathematics are fairly obvious. If he is to help his pupils develop ability in mathematics, he must develop abilities in the three modes through which mathematical thought is transmitted. He must see that students develop a mathematical vocabulary which is solidly grounded in common meanings arising from common experiences. He must also see that the usage rules for this vocabulary are logical, appropriate, and consistent, requiring more than mere memorization of terms. Careful development of ideas and the extended use of these ideas are essential. Few mathematics teachers require their students to write short themes or essays explaining mathematical relationships, but this practice would be a beneficial one in developing semantic ability in mathematics.

Most classroom time in mathematics is spent dealing with symbols in a manipulative sense. The algorithms of arithmetic have most of their origin in the symbolic conventions used in writing numerals. For example, addition requires a "carrying" process not because carrying is basic to addition, but because regrouping is necessary in the place value system we use in writing numerals. These rules for symbol manipulation, which permeate so much of arithmetic, extend to the various areas of algebra. Most students even see geometry as a kind of complex symbol manipulation. Ask any adult what he remembers about proof in geometry, and the chances are very good that he will answer, "Well, it has something to do with dividing the page into two columns so that every time you write something in the lefthand column you have to write a reason in the righthand column." What is being remembered, of course, is not inductive or deductive methods and strategies but the symbolic conventions used to organize these methods.

Manipulating symbols is not a situation from which we can escape entirely. Much of mathematics does reside in its symbolic systems, and it is entirely proper for mathematics teachers to see the development of symbolic ability in mathematics as a major goal. Nevertheless, it is highly doubtful that much of the symbolic manipulation in mathematics classes really develops this ability. The strength of symbols depends upon the establishment of a kind of isomorphic mapping between the realm of ideas and the realm of symbols. In translating ideas to symbols, manipulating symbols, and translating the result back to the corresponding idea, we tend to emphasize the middle step of this process — the manipulation — and ignore the vital translations at the beginning and end. True ability in symbolic mathematics resides, however, in the complete process, and it is this entire process that must be emphasized in mathematics classrooms. Hopefully, this distinction is one of the most meaningful that can be made between older mathematics curricula and the "new mathematics." A mathematics textbook from the 1940's is full of examples — and little else. The emphasis is on manipulation through a "do as I do" approach. Sample problems are worked with each step clearly marked, but seldom explained. In contrast, the first textbooks of the University of Illinois Committee on School Mathematics (UICSM) and the School Mathematics Study Group (SMSG), among others, were highly verbal. In fact, a common complaint was that the texts were so verbal that students could not read them. But in large measure, this verbiage was an attempt to emphasize the translation processes so important in developing symbolic ability.

However, those translation processes need not always be carried out verbally. One can translate by focusing on objects or images which have a property that is an embodiment of the mathematical idea to be considered. The objects or images can be handled by means of figures — symbols which include the basic elements of the objects being represented. Figures can then be gradually transformed to symbols as logical rules for operations are developed. This process is crucial for elementary school mathematics. Although it is not a prerequisite procedure for higher level mathematics, we must not assume that figural ability plays no role in the mathematical ability of a research mathematician. The "doodling" involved in problem solving is really a process of figural reasoning.

The "doodles" are usually incomplete figures distilled to only essential features. Because they are highly distilled, they are usually incomprehensible to the casual observer. But to the problem solver they clearly play a figural role. Diagrams are important not only in geometry, but in all branches of mathematics. Teachers should require students to draw pictures which illustrate mathematical ideas.

Not only does ability in mathematics imply ability in the modes in which mathematics is communicated, but as we have seen, these modes actually reinforce each other. Thus, to talk about semantic mathematics without referring to symbolic mathematics and figural mathematics is impossible. Nevertheless, school mathematics programs commonly emphasize some modes more than others. A clear understanding of the distinction between semantic, symbolic, and figural content is the first step toward restoring balance to the development of mathematical abilities.

Product Abilities in Mathematics

We have seen that different abilities for processing information according to the mode in which it is transmitted are reflected in differences in mathematical ability. However, it would be surprising if we did not also find differences in ability due to the kind of information being transmitted in any particular mode. These differences are implied when we ask if there is an ability to "do" mathematics as opposed to an ability to "do" English, etc. The product dimension of Guilford's model distinguishes between basic differences in the nature of the information received. The product categories are not as simple as mathematics, science, history, language, etc. Instead they are basic categories of information which might exist within any subject-matter discipline. Guilford defines the six product categories this way:

Information can be conceived in the form of *units* — things, segregated wholes, figures on grounds, or "chunks." Units are things to which nouns are normally applied. *Class,* as a kind of product of information, is near to the common meaning of the term. A class is a set of objects with one or more common properties; but it is more than a set, for a class idea is involved.

A *relation* is some kind of connection between two things, a kind of bridge or connecting link having its own character. Propositions commonly express relation ideas, alone or with other terms, such as the expressions "married to," "son of," and "harder than." *Systems* are complexes, patterns, or organizations of interdependent or interacting parts, such as a verbally stated arithmetic problem, an outline, a mathematical equation, or a plan or program. *Transformations* are changes, revisions, redefinitions, or modifications by which any product of information in one state goes over into another state. Although there is an implication of process in this definition, a transformation can be an object of cognition or thought like any other product. The part of speech that we ordinarily apply to a transformation is a participle, a verb in noun form, such as shrinking, inverting, or reddening.

Finally, an *implication* is something expected, anticipated, or predicted from given information. Behaviorists who admitted the concept of "expectation" or "anticipation" to their lexicons have been talking about much the same idea. Any information that comes along very promptly suggests something else. One thing suggesting another involves a product of implication. Of all the six kinds of products, implication is closest to the ancient concept of association. But something more is involved in the concept. It is not that one thing merely follows another but that the two have some intimate way of being connected. This does not make an implication the same as a relation, for a relation is more specifiable and verbalizable. (2:63-64)

This analysis of the basic categories or products of information could serve as a basic analysis of the nature of mathematics with only minor modifications. Guilford's *units* of information are mirrored by the basic *elements* of a mathematical system: the numbers of arithmetic, the variables of algebra, the points, lines, and planes of geometry. Informational units may be more complex, of course, and so may mathematical elements. For example, if the system we are studying is modular arithmetic of modulus 4, then the elements of the system are four equivalence classes. In a similar manner, if the discussion involves different kinds of mathematical transformations, each treated as a single idea, then the information unit would be a mathematical transformation. Guilford says, "Other kinds of products can become units, as they acquire 'thing' character. A relation can become an object of interest as an event; so can a class or a system." (2:239)

The classes of information correspond to well-defined mathematical sets, with one exception. For a set to be well-defined the definition must enable one to identify the elements of the set. This identification can be done by specifying the basic common characteristics of all the elements. For example, we could define the set of even numbers, E, as the set of all whole numbers divisible by two. These common characteristics (whole numbers, divisible by two) establish the "class idea" to which Guilford refers. In contrast, we can also define a mathematical set by simply specifying all the elements of that set. If we specify the set B as

$$B = \{7, \quad 13, \quad 19, \quad 27\}$$

it, too, is well defined. But sets defined in this manner may or may not possess a "class idea." Often, whether or not there is a class idea involved cannot be determined by simple examination.

The *relation* of information products is exactly the same as the relation of mathematics: a set of ordered pairs. However, as one might expect, the relative emphasis is different. In mathematics we are often satisfied to determine only if the relation exists; with Guilford's products, the nature of the relationship is of primary importance.

Guilford's *systems* are considerably broader than the systems of mathematics. Mathematicians would not recognize a "verbally stated arithmetic problem" as a system. Nevertheless, there is a common bond. An arithmetic problem is solved by creating a mathematical model of the problem within a particular mathematical system. In this sense, the appropriate system of mathematics does become the most important feature of the problem. It is in both the systems of information products and the systems of mathematics that we reach the real payoff and the first sense of completeness. An information system is what information is all about, just as a mathematical system is what mathematics is all about. Systems provide the structure of mathematics.

Transformations lie at the heart of mathematics. The transformations of information products become the *functions* and *operations* of mathematics. Transformations exist not only as specified operations within mathematical systems, but as mappings from one mathematical system to another. Just as transformations are basic to mathematics, transformations are basic to knowledge of information in general.

Faculty psychologists believed that the study of certain disciplines increased mental abilities. Mathematics was supposed to increase the general ability for logical thinking. The logical foundations upon which mathematics is based correspond closely to the implication category of the products dimension. The formal "if . . . then . . ." implication of mathematical logic is reflected in the implications of everyday activities. Mathematical implication is clearly subsumed wthin the implication category of the product dimension.

The product dimension of Guilford's model of the intellect divides knowledge into psychological categories. We have seen that mathematics can be divided into closely corresponding logical categories. Is there a causal relationship? Do the categories correspond because mathematics is an intellectual activity, or do they correspond because in overall perspective intelligence is basically mathematical? These questions will be explored in more detail in later units.

Process Abilities in Mathematics (Operations)

Guilford's operations dimension represents what we do with information once we receive it. In this sense it identifies basic ways in which the human organism processes the information it receives. The ability to process information in different ways varies from individual to individual. How is it reflected in mathematical abilities?

The five operations categories of Guilford's model — cognition, memory, convergent production, divergent production, and evaluation — represent ways of thinking about or acting on information. However, the categories can also be considered in terms of the complexity of thought necessary to operate in each way. If they are viewed in those terms, some rearranging would be in order. Memory becomes the simplest level. Guilford defines memory as "retention or storage, with some availability, of information in the same form in which it was committed to storage and in connection with the same cues with which it was learned." (2:211) This definition implies an absolute minimum of processing of the original information. Recall, not processing, is the important factor in this category, which is precisely what is important in *computation* in mathematics, when a specific algorithm is carried out. We do not mean to include the process of choosing which algorithm to apply — obviously a more complicated procedure than simple recall. But once the type of computation to be performed has been specified, carrying out the algorithm involves recall of the steps "in the same form in which (they were) committed to storage and in connection with the same cues with which (they) were learned." Differing abilities in *memory* imply different abilities in *computation* in mathematics. There may be other minor abilities in mathematics which require no more than recall, but the area of computation is certainly the most significant area.

Cognition becomes the next level in the operations dimension. Guilford defines cognition as "awareness, immediate discovery or rediscovery, or recognition of information in various forms; comprehension or understanding." (2:203) Certainly awareness or recognition depends upon memory; but more than recall, it requires bringing stored information to bear upon the acquistion of new information. Cognition is probably the broadest of the five categories in this dimension. Cognition includes recognizing the problem, "find the cost of 12 apples if one apple costs 6¢" as a "multiplication problem." Cognition includes looking at the multiplication facts of nine ($1 \times 9 = 9$, $2 \times 9 = 18$, $3 \times 9 = 27$, $4 \times 9 = 36$, etc.) and realizing that the sums of the digits in each product is also nine. Cognition includes following the work of the mathematics teacher on the blackboard to the extent that one understands the pattern formed by the steps and can suggest what the next step might be if that pattern were continued.

In mathematics, then, cognition refers to a kind of basic *comprehension*. The student understands what is being communicated. Nevertheless, cognition or comprehension implies a rather low level of understanding. The student may be aware of the pattern in the products of nine, for example, but may not be able to relate this result to more basic number properties. This relating often calls for the generation of new information by the student in the form of connecting links. Such generation of information from given information in the sequence of logical necessities defines the category of *convergent production*. Convergent production suggests that the thinking focuses towards one result, one final correct answer, which is clearly the realm of deductive thinking in mathematics.

In contrast to convergent production is *divergent production*. New information is generated from old, but the "emphasis is upon variety and quantity of output from the same source; likely to involve transfer." (2:213) Because the emphasis in divergent thinking is upon all possible combinations rather than the single focus of convergent thinking, to place divergent production at a higher cognitive level than convergent production is reasonable.

In the mathematics classroom, these two ability categories are usually observed comingled together as *problem-solving* ability. Problem solving in mathematics often begins with divergent thinking. The mathematician seeks clues to the structure of the problem through a kind of intelligent guessing procedure. This trial and error procedure requires divergent production to generate new combinations and possibilities to be tried. The mathematician does not solve problems by random guesses, but he brings forth his store of previous knowledge to generate attempts based upon cognizance of the procedures of *application* of mathematical models and the careful analysis of the problem at hand.

The outcome of this divergent thinking phase of problem solving is new information gathered from the trials. If the mathematician is able to put this information together in a related way to form a *synthesis*, then the problem solving proceeds to a phase of convergent thinking where the information is focused almost deductively towards a solution. (If a synthesis cannot be achieved from the divergent thinking phase, the solution is stymied, and the problem solver fails.) Of course many problems are stated in such a way that their structure is almost immediately apparent.

In these cases the divergent thinking phase is minimal or non-existent. We must *not* assume that because divergent thinking can be minimal or that because it usually occurs first in the process of problem solving, it is somehow simpler than convergent production. Because of its emphasis on variety, divergent thinking represents a higher, more complex cognitive level than convergent thinking.

The last category in the operations dimension is *evaluation*. Guilford defines evaluation as "a process of comparing a product of information with known information according to logical criteria, reaching a decision concerning criterion satisfaction." (2:217) Perhaps it seems strange that this category is the highest, most complex cognitive level. However, if one views evaluation as decision making, it becomes apparent that in this category we integrate all of the abilities represented in the previous levels. Our evaluation means more than simply judging something to be good or bad, liked or disliked. In mathematics, evaluation implies the ability to integrate knowledge and skills from a wide variety of areas and to judge their relative merits against a set of criteria. In this sense, evaluation is the highest kind of creativity brought to bear on the problem level — the problem of *production* of new useful knowledge in mathematics. Both the newness and the usefulness of this knowledge must be understood in terms of the learner. A creative discovery might be at the level of *evaluation* for a third grader, yet be at no more than the *memory* level for a research mathematician. That the discovery is new information for the third grader *and* that he was able to judge it against the criteria of his previous knowledge as being valid, related to mathematics, and useful to him in his mathematical endeavors is important. This kind of production is a discovery quite different from that of mere recognition or awareness. It usually comes through intensive problem solving. The solution to the problem is evaluated in light of previously known mathematics. If the solution can be integrated as an extension of the previous mathematics, it becomes accepted as new mathematics. Thus, the key to the *production* process in mathematics is not discovery but *evaluation*.

Interestingly enough, the ability to operate at a production or evaluation level is one of the prerequisites to responsive teaching. The teacher who is responsive to the ideas of his students cannot always know what the final outcome of those ideas may be. But he can evaluate this final outcome against the criteria of the mathematics he does know. If the teacher becomes confident in his abilities to evaluate, he becomes able to be responsive to the explorations of his students. In contrast, the teacher who cannot judge whether new ideas are mathematically useful must suppress the divergent explorations of his students and must stay with what he already knows to be useful. As a result he lacks the ability to be flexible which is so characteristic of responsive teaching.

We can contrast Guilford's operations abilities for processing information with the corresponding abilities in mathematics in the following chart.

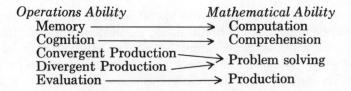

Figure 8. Guilford's Operations and Mathematics Ability.

These levels not only identify different ways in which information is processed, but they form a hierarchy of levels of complexity of cognitive functioning. They suggest that mathematics teachers should broaden their expectations of student behavior in mathematics. The mathematics teacher who stresses only computation in his classes is requiring only the lowest cognitive functioning of his students. Most problems on mathematics tests are intensively focused on obtaining a single answer, sometimes by a single "correct" process. This kind of testing requires no more than cognition plus a minimum of convergent production. We seldom ask students for divergent production by means of questions which ask for as many different ways to solve a problem as a student can think of. We need to consider carefully the process categories when establishing goals and objectives for mathematics.

A committee of college examiners, established in 1948, developed a theoretical framework for classifying the goals and objectives of the general educational process. The committee was interested in such a framework since educational objectives provide the basis for curriculum building, testing, and educational research. Working independently of Guilford, the committee developed a *Taxonomy of Educational Objectives* for the cognitive level which bears a striking similarity to the categories of Guilford's operation dimension (1). The taxonomy, often called "Bloom's taxonomy" after its editor, Benjamin Bloom, contains six levels. They are:*

1:00 Knowledge

Knowledge, as a taxonomy category, simply means that a student can show that he knows an idea. He need only remember the idea in a form very close to the way it was originally encountered. This category emphasizes memory through either recall or recognition. Although remembering may be simply rote recall, it may also involve complex psychological processes. If one thinks of the mind as a file, then the problem of recall is to find in the problem or task situation the appropriate signals, cues, and clues which will most effectively bring out whatever knowledge is filed or stored.

2:00 Comprehension

Comprehension represents the lowest level of understanding — the understanding of the literal message contained in a communication. To show that he comprehends, the student may be required to put the communication into other terms or into parallel forms. He may need to make inferences, summarizations, or generalizations about the ideas communicated. However, he will not necessarily relate the ideas to other material or see the fullest implications of the ideas.

3:00 Application

Application implies the ability to use an appropriate idea or abstraction in a problem situation. In an application the problem should be new to the student and there should be no additional hint as to which concept to use or how to use it. It is these restrictions that make application a higher-order behavior than comprehension. "A demonstration of 'Comprehension' shows that the student *can* use the abstraction when its use is specified. A demonstration of 'Application' shows that he *will* use it correctly, given an appropriate situation in which no mode of solution is specified." (1:120)

4:00 Analysis

Analysis emphasizes the ability to break material into its constituent parts and to detect the relationship and organization of these parts. Analysis may be an aid to fuller comprehension if it clarifies the communication and indicates how the communication is organized. It may also be a prelude to evaluation if it shows how the communication manages to convey its effects or purposes.

5:00 Synthesis

Synthesis is defined as the putting together of elements and parts to form a whole. This process involves working with pieces, parts, and elements from many sources and arranging them to form a pattern or structure not clearly present before. The emphasis upon uniqueness and originality calls for creative behavior on the part of the learner.

6:00 Evaluation

Evaluation is the making of judgments about the value of material and methods for given purposes. It involves the use of criteria which may be determined by the student or given to him by another source. Evaluation is placed last in the taxonomy because it requires all the other categories of behavior to some extent. It is not necessarily the last step in thinking or problem solving, however; for it may trigger new knowledge, comprehension, and application or a new analysis and synthesis.

The cognitive levels in Bloom's *Taxonomy* relate to our discussion as indicated in the following chart:

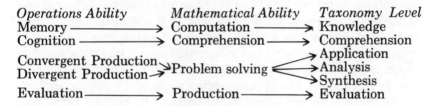

Figure 9. Guilford's Operations, Bloom's Taxonomy, and Mathematics Ability.

As the chart shows, the primary difference between Guilford's operation abilities and Bloom's cognitive levels comes in the area of problem solving. These differences really reflect basic differences in the methods used to generate the two cognitive-level schemes. Guilford generates intellectual factors out of test data. Bloom, on the other hand, looks also at basic objectives underlying the content we teach. The categories of application, analysis, and synthesis make a more detailed examination of the procedures used in problem solving than the more global convergent and divergent thinking categories. Every problem solution in mathematics comes from an *application* of more basic facts and principles. This kind of application is far broader than what one usually means when speaking of "applied mathematics." It focuses not so much upon practical uses for mathematics as it does upon useful relationships between mathematical principles. Application first asks, "can this same mathematical idea be found in another setting?" For example, we commonly encounter the ratio pi in the study of circles. But later we find this value also useful in trigonometry because defining trigonometric functions in terms of circles is easy. We are able to apply the idea of the ratio pi in a new setting. In this sense *application* is used as a category in Bloom's taxonomy. Problem solving begins in the application of old ideas to new situations.

Of course, equating problem solving with "mere" application is not enough. If that were the case, teaching children how to solve mathematical problems would be easy. We would simply catalog all the basic application patterns and teach them the catalog. Some textbooks do suggest that the authors have attempted cataloging. Older algebra texts often classified word problems in tight little categories. Students studied age problems, mixture problems, rate problems, etc. The difficulty with texts of this sort is that there is very little structure between the categories. As a result one gets bogged down in detail. There is no way to tell within a given category which category should logically follow. As a result, problem solving degenerates to a "bag of tricks." What is missing, of course, is *analysis*. What parts of the problem are essential information? What relative roles do these parts play in the basic problem? Often these analysis questions can be pursued by changing the number values given in a problem to see if certain values make the problem meaningless. This process might be a kind of divergent-thinking one. In other cases, the problem solver might carry out the analysis by relating the variables to another problem he already knows how to solve. This might be a convergent-thinking process. But even in this case, the solver is looking beyond the situational category of the problem to more fundamental patterns and relationships.

Finally, having found ways to apply familiar principles to new situations and having analyzed the new situations into their basic components, the problem solver must begin to put all the pieces together to generate a final solution. This process of fitting old components together in new ways is basic to the *synthesis* level. Like application and analysis, synthesis may also call for convergent or divergent thinking. The components may be fitted together in a deductive fashion, or they may be combined through a creative kind of trial and error process. When people write down problem solutions, they almost always organize them in a logical deductive fashion. This organization makes the solution easy for a reader to follow, but it does not mean that synthesis requires only convergent thinking. Problem solutions are often synthesized by the free-thinking which characterizes divergent production.

Who is right, Guilford or Bloom? At this point in time we cannot say. Perhaps both are wrong if we must judge in an absolute sense. Perhaps future studies will suggest a different way to look at the structure of problem solving abilities. Nevertheless, until that time, both Guilford and Bloom provide useful ways of looking at the organization of abilities in problem solving. The development of problem-solving ability is the one objective for school mathematics that most educators can agree on.

Modes of Mathematical Explanation

It is easy to see how Guilford's structure of the intellect can be used by the mathematics education theorist and researcher. What does it mean to the classroom mathematics teacher? If a teacher believes that the abilities of his students vary according to the structure set forth by Guilford, is the belief reflected in his classroom teaching? What would a classroom observer see in watching a teacher that would identify him as a "Guilford believer"?

First, such a teacher should have an amazing variety of teaching approaches. A teacher who believes in the Guilford model stands before a class of thirty pupils with fear and trembling! But his fear is purposeful. He has counted all the cells in the model, and he realizes that a single pupil represents as many as 120 different facets of abilities. When he faces his class he faces not just 30 different degrees of ability which will be brought to bear on his mathematics lesson, but as many as 30×120, or 3,600 different degrees of ability to learn. If he is to present a lesson in mathematics which can be understood by his pupils, he must be ready to adapt to the wide variety and range of abilities in his classroom. Certainly one implication of the Guilford model is a call for variety and diversity in lesson presentation.

Nowhere is this call for variety and diversity more apparent than in the explanations the teacher gives of mathematical content. He must realize at once that a single, once-over explanation of a mathematics concept or process is not going to suffice for any class. Some children will be able to learn best from content that is presented in a figural or pictorial mode. Others will learn easily if the teaching presents the material in a conversational or semantic mode. Yet others will need to connect the content to symbolic representation to learn efficiently. If the teacher believes in the Guilford model, he realizes that a child who raises his hand and says "I don't understand" does not need more of the same explanation, but he needs different explanations which rely on different content modes.

In many ways, the need for a wide variety and diversity of explanations in teaching should be obvious. If a single explanation were always sufficient for learning, teachers would have been out of business centuries ago! If a single explanation is all that is needed for good teaching, there is certainly a much more efficient way to provide it than hiring a teacher to stand in front of a class. The textbook provides a single explanation of each important idea in the course. If that explanation alone were sufficient, teaching would have evolved into handing each student the appropriate textbook and leaving him alone! Of course, a few students can learn this way. But the fact that these students are very rare is indirect evidence that there is a wide variation in abilities in the categories along Guilford's *content* dimension.

That the teacher who believes in the Guilford model provides a variety and diversity of explanations hardly explains how he knows which explanations to give. If he knew the intelligence profile for each student in his class, he might begin to tailor his lessons to their specific abilities. Such tailoring is not an easy thing to know how to do, and it represents a major area of current interest in educational research known as aptitude-treatment interaction. After considerable study educators still do not know precisely how to construct teaching treatments which will interact with a student's abilities to produce observable differences in learning. Until this capability is developed, teachers can insure that students get a variety of explanations. Nevertheless, a teacher need not flounder around aimlessly in search of changing explanations. The structure of the intellect can suggest ways to look for a variety of explanations.

Remember that the dimension called *content* has three divisions that are important in mathematics. (We shall not deal with the behavioral category.) These divisions are semantic, symbolic, and figural. Many, many concepts in mathematics can be represented in each of these three ways. Consider numbers, for example. We can deal with numbers on a purely semantic (verbal) basis in mathematics. We can also deal with numbers through a use of the numerals, or symbols which represent them. But in many cases, we can deal most effectively with numbers through a figural representation. The best figural representation of numbers is, of course, that of the number line. The number line presents a powerful tool for constructing explanations which do not depend upon symbol-manipulation or verbal reasoning.

The analysis and identification of a teacher's explanations as verbal (semantic) or symbolic or graphic (figural) is not as easy as it might first seem. Even the teacher who uses the number line does a lot of talking in the course of his explanation. If we look merely for the presence of these modes, then every explanation is verbal (except for those rare teachers who know how to use nonverbal techniques). What we must look for is not the mere existence of any particular mode but the use and importance of a mode within an explanation. If a teacher talks about symbols which are being written (or have been written) on a blackboard or projection transparency, then we may usually assume that his explanation is symbolic. Perhaps as his explanation unfolds, that he is not talking about the symbols as much as the way the symbols are moved around spatially or result in a spatial picture becomes apparent. Then his explanation is not symbolic but graphic. A good example of an explanation which *seems* symbolic, but is really a graphic explanation is the "happy face" method for remembering binomial multiplication. This explanation suggests that the student can remember the result of the binomial multiplication pattern by drawing a "happy face" on the indicated product.

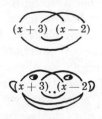

Figure 10. Binomial Multiplication Pattern.

Not very good mathematics, but often an effective memory aid! We must simply remember that the product of the two terms at the ends of a curve are a partial product in the final answer. Despite all the writing, the important part of this explanation is not the symbols but their relative positions. Write the binomial factors in any other way, and the "happy face" explanation does not work. It is the importance of the spatial orientation and the pictorial quality that makes this a graphic explanation.

Of course, one might argue that all symbols involve spatial orientation. Certainly 23 is not the same as 32, and $5x$ is not the same as 5^x. But we prefer to view these differences as differences of combination and order rather than differences of major spatial orientation. Certainly they do not carry the pictorial quality evident in the "happy face" or in the number line or in other graphs.

When we try to construct a graphic explanation in mathematics then, we must not expect to completely avoid the use of words or symbols. But the basic intent of the explanation must be to picture the idea. Similarly, a symbolic explanation may use words and employ spatial orientation, but it must primarily suggest the symbolic representation of ideas and the rules for combining and working with these symbols. Only in verbal explanations should we look first for the absence of the other two modes.

Mathematical concepts may be explained emphasizing each of the three modes — verbal, symbolical, and figural. You may want to try your hand at alternate representations as you read the examples.

1. Place Value

Suppose we begin with the place value representation of the number system we commonly use. This representation is a symbolic concept, and we are undertaking the explanation of a symbolic system. Nevertheless, we can construct an explanation that is primarily verbal. If each number were represented by a single symbol, we would have a lot of symbols to learn. An alternative would be to use a limited number of symbols so that each particular one could stand for several different numbers according to some additional kind of key. For example, we might use color as an additional key so that a black 2 and a red 2 represented different numbers. If we bunched quantities in some way, we might have the black 2 represent two single elements (the basic concept of twoness) while the red 2 stood for two of the larger bunches. We have yet to decide, of course, how large to make a bunch. In practice this decision is an historical one. In theory it depends upon how many single elements we can remember before needing to group (remember that the basic memory unit from our investigation was about 7, plus or minus 2), the number of different symbols we can conveniently construct, and the presence of naturally bunched objects (fingers on the hands, leaves on a clover stalk, etc.)

Our Hindu-Arabic scheme is built on ten. We agree to use ten basic symbols or digits, zero through nine. We use the digits to represent different bunches through a key system built not on color, as in our example, but on the position of the digit in a composite symbol or numeral. By agreement, numerals are built from right to left. Digits in the first position (farthest right) represent the number of single elements; those in the second position immediately to the left represent the number of bunches of ten elements; those in the third position represent bunches of ten times ten, or one hundred elements.

For example, to represent the number one hundred forty-six, we must decide the largest accepted bunch in our system that it contains. Our number contains one one hundred, so it is represented by writing the digit, one, in the third position. Forty-six remains. Four bunches of ten are contained in forty-six, so four is written in the second position. Six remains, and the digit, six, is written in the first position. The resulting numeral is 146.

The digit zero, used to represent no single elements or bunches, requires an additional convention. There are no grouping bunches larger than the hundreds group in our particular number. Should they be represented by writing zero in their positions? Why not write 000000000000000146? Can you express the convention we accept for the use of zero? Is zero always a placeholder?

How do first graders manage to learn the number system at all? There is a kind of insanity in giving a purely verbal or semantic explanation of numerals. How much of your ability to follow this explanation was due to the fact that you had previous knowledge of what was being explained? At the very least, we need to see the symbols that are being explained. We could enhance the preceding explanation by writing:

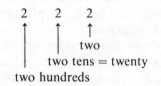

hence, one hundred forty-six is written:

How can we represent the same idea in figures? Many textbooks show bundles of sticks. A more distinct visual representation has been suggested by Z. P. Dienes. Dienes uses a small wooden cube as a basic object or unit. The tens units is represented by a rectangular wooden stick whose cross section looks like the cube, but whose length is ten times the basic dimension of the cube.

The hundreds unit is represented by a block of wood as thick as the cube but ten times longer and ten times wider.

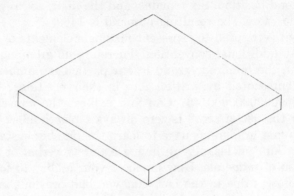

Figure 11. Place Value in Figures.

Children are asked to use the wooden pieces to represent one hundred forty-six cubes. Most children quickly see that the best representation (to limit the handling of separate pieces) is one block, four sticks and six cubes.

The principal of these blocks does not depend upon a base of 10, of course. We could easily construct a base 3 system by making sticks whose length is three times that of a cube, and blocks whose dimensions are 1x3x3. For this reason, the system is known as Multibase Arithmetic Blocks (MAB). Such blocks are very useful in developing carrying and regrouping techniques for addition and subtraction in various bases.

2. Multiplication of Binomials

What does *binomial* mean in algebra? Too often definitions are limited to explanations of symbols: "A *binomial* is an expression of the form $(a+b)$." The definition really needed is: a number which has two additive components. This definition means that the number is considered in two ways: as a single number or as the sum of two parts. To multiply two such numbers together, consider one number as a sum which is to be multiplied by a single number and apply the fact that multiplication is distributive over addition.

For the product $(a+2) \times (a+3)$ this process yields $a(a+3) + 2(a+3)$. Then instead of considering the second number $(a+3)$ as a single entity we now consider it as the sum of two components and again apply the distributive law, which yields $a^2 + 3a + 2a + 6$ or $a^2 + 5a + 6$.

We can develop an alternate explanation which emphasizes symbols by relating multiplication of binomials to symbolic processes the student already knows. For example, we can view the operations of algebra as natural extensions and generalizations of the operations of arithmetic. Therefore, there should be some analogy between the multiplication.

$$\begin{array}{r} a+2 \\ a+3 \\ \hline \end{array} \quad \text{and the multiplication} \quad \begin{array}{r} 32 \\ 41 \\ \hline \end{array}$$

In fact, we can separate 32 and 41 into components using the place value concept as our cue. We have:

$$\begin{array}{r} 32 \\ 41 \\ \hline \end{array} \longrightarrow \begin{array}{r} 30+2 \\ 40+1 \\ \hline \end{array} \longrightarrow \begin{array}{r} a+b \\ c+d \\ \hline \end{array}$$

The answer to the first problem is:

$$\begin{array}{r} 32 \\ 41 \\ \hline 32 \\ 128 \\ \hline 1312 \end{array}$$

Once we realize that the second partial product is indented one space to the left to account for the fact that we are really multiplying by the number 40, we can see that in expanded form the problem solution looks like

$$\begin{array}{r} 30+2 \\ 40+1 \\ \hline 30+2 \\ 1200+\ 80 \\ \hline 1200+110+2 \end{array}$$

Note that again we have used the distributive law twice. Then by analogy, the algebra solution — which has no place values and, hence, should follow the expanded notation — should be

$$\begin{array}{r} a+2 \\ a+3 \\ \hline 3a+6 \\ a^2+2a \\ \hline a^2+5a+6 \end{array}$$

Finally we may explain the multiplication of $(a+2)(a+3)$ figurally. The area measure of a rectangular region is equal to the product of the length measure of two perpendicular sides. We can construct a rectangle which is $(a+2)$ by $(a+3)$.

When the measures of the sides are extended we see that the total region is made up of three types of components. One is the square whose measure is a^2. The second is the square whose measure is $1^2=1$. The last is the rectangle whose measure is $1 \times a = a$. To find the measure of the region, we have only to count the component regions and sum their measures. We have a^2+5a+6. Since the measure of the total region is also $(a+2)(a+3)$, we have $(a+2)(a+3) = a^2+5a+6$.

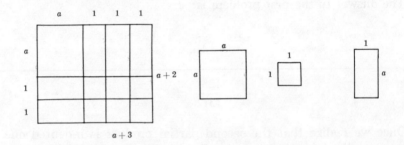

Figure 12. Binomial Multiplication.

For classroom effectiveness, students can build products of different pairs of binomials from a set of several copies of the three basic regions which have been cut from cardboard. The length chosen for a should not be a multiple of the length chosen for 1.

3. The Irrationality of $\sqrt{2}$

The most popular proof of the irrationality of the square root of two can be given in a form that is essentially verbal. Assume that the square root of two is a fraction expressed in lowest terms. Then the square of this fraction is the number two, which implies that the numerator of the squared fraction is even. If a perfect square is even, then it must also be divisible by four. (Why should this be so? Think of numbers that are perfect squares — can you find an exception?)

If the value of the squared fraction is only two, and the numerator is divisible by four, then the denominator must also contain at least one factor of two. But if both the numerator and the denominator are divisible by two, the fraction is not in lowest terms.

This contradicts the original assumption, and we can only conclude that the square root of two cannot be expressed as a fraction.

Now consider a more symbolic proof of the irrationality of $\sqrt{2}$.

1. Assume that $\sqrt{2} = p/q$ in lowest terms.

2. $1/\sqrt{2} = \dfrac{\sqrt{2}}{\sqrt{2}\sqrt{2}} = \frac{1}{2}\sqrt{2}$; therefore, $q/p = \frac{1}{2}(p/q)$.

3. This means that q/p and $p/2q$ are in the same equivalence class.

4. Since q/p is in lowest terms, there must be an integer, k, so that $p = kq$ (and $2q = kp$).

5. Then $p/q = k$, that is $\sqrt{2}$ is an integer.

6. But $1 < \sqrt{2}$ since $1^2 < 2$, and $2 > \sqrt{2}$ since $2^2 > 2$.

7. Therefore $\sqrt{2}$ cannot be an integer. Contradiction.

These two proofs of the irrationality of $\sqrt{2}$ tend to show the pitfalls of relying too heavily on either a verbal explanation or a symbolic explanation. The verbal explanation cries out for symbols when we have to consider the numerators and denominators of the assumed fraction and its square separately. At the same time, the symbolic proof demands more verbal explanation. Step 4 particularly needs more expansion. What we are doing in step 4 is referring to a previous theorem about equivalent fractions. We need to look at an example of an equivalence class of fractions in order to develop an intuitive understanding of this theorem. Consider fractions that are equivalent to $\frac{2}{3}$. Examples would be $\frac{4}{6}$, $\frac{6}{9}$, $\frac{8}{12}$, $\frac{10}{15}$, $\frac{12}{18}$, $\frac{14}{21}$, $\frac{16}{24}$. The simplest way to generate members of this class is to pick an integer, multiply it first times the numerator of the lowest-terms fraction in the class to obtain a new numerator; multiply it times the denominator of the lowest-terms fraction to obtain a new denominator. This operation generates a new fraction belonging to the equivalence class. Step 4 maintains that we can always reverse the procedure. This verbal addition to our symbolic explanation does not destroy the basic symbolic intent of the explanation, but it does point out the benefit of judiciously mixing modes of explanation.

For a figural demonstration of the irrationality of the $\sqrt{2}$ we can use an argument which utilizes geometric constructions. Construct a right triangle ABC whose legs each measure one unit. By the Pythagorean theorem the measure of the hypotenuse (AC) is $\sqrt{2}$ units. Assume that there is a segment of length l/q (where q is an integer) such that both the legs and the hypotenuse are *commensurable*. (A line segment is commensurable if there exists an integer n such that the segment formed by n measure-segments is congruent to the original segment. The legs of the triangle are commensurable by the measure-segment l/q since q of them would form a segment of length l. Assuming that $\sqrt{2}$ is also commensurable is equivalent to assuming that there exists an integer p such that a line segment made up of $p \times l/q$ measure-segments is congruent to the hypotenuse. That is, that $p/q = \sqrt{2}$. Hence, the assumption of commensurability is equivalent to assuming that $\sqrt{2}$ is rational.)

Begin by constructing AD on AC so that AD = AB. Since AC and AB are both commensurable, DC must also be commensurable. (Its measure is pq, and pq is an integer.)

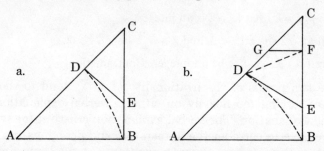

Construct DE ⊥ AC at point D. BE ≅ DE and DE ≅ DC. (Can you supply the proofs?) Since BC and BE are both commensurable, CE is commensurable. Furthermore, CE < CB, since CE + EB = CB.

Construct EF on EC so that EF ≅ DE. Since EC and EF are commensurable, CF must be commensurable, and CF < CE < CB. Notice that △CFG is similar to our original △ABC. By repeating this sequence of constructions, we get the sequence of shorter and shorter commensurable segments CM < CJ < CI < CF < CE < CB shown here:

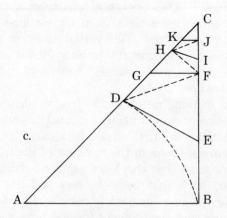

Figure 13. The Irrationality of $\sqrt{2}$.

Except for matters of precision in making the construction, there is no reason why this procedure must ever terminate. We can construct a commensurable segment as small as we want. In particular, we can construct a commensurable segment smaller than our measure-segment l/q! The only way out of this paradox is to deny our assumption that both l and $\sqrt{2}$ are commensurable. Therefore, $\sqrt{2}$ must be irrational.

Can all mathematical concepts be explained in semantic, symbolic, and figural modes? To insist that they could would certainly be pressing the case too far. There are some topics which are primarily symbolic, and others which are primarily figural. But the teacher who accepts the Guilford model should always be on the search for alternative modes of explanation.

1-5: Study Module

For Further Investigation and Discussion

1. Guilford's intellectual factors depend upon our ability to construct the right kinds of test situations and test items. How do the items of commercially available standardized tests reflect the range of intellectual factors that Guilford has found? Borrow a copy of a commercial achievement test in mathematics or a copy of the quantitative section of an intelligence test. Decide the category of each item for content, product, and operation. Of the 120 possible combinations, how many are represented?

2. It could be argued that the structure of intellect which Guilford has identified does not represent the human mind at all and that those factors only appear because of the way test items are normally constructed. Argue for or against this position.

3. Choose a chapter in a mathematics textbook. Identify the sections which are primarily explanatory. Decide what mode of explanation each section emphasizes. Can you characterize the text as verbal, symbolic, or graphic?

4. Repeat the analysis in item 3 for a similar chapter in a similar text published in the 1940's. Do the same for a textbook published around the turn of the century. Do you find differences in modes of explanation? If you divide work up among class members, you may be able to trace a particular chapter or topic through several years. Can you find trends or "fads" in mathematical explanation?

For Lesson Planning

Give brief symbolic and graphic explanations or descriptions of the following mathematical ideas or statements.

1. Addition is commutative.
2. The relationship "greater than" is transitive.
3. The sum of two odd numbers is an even number.
4. The average of two numbers is the value midway between the two numbers. Explain "midway between the two numbers," then show that this figure is the same as half the sum of the two numbers.
5. The perimeter of a rectangle is double the sum of its length and width.
6. $(x+6)(x-2) = x^2 + 4x - 12$.
7. Four halves of a number is the same as twice the number.
8. Twice the sum of two numbers is the same as the sum of twice the first number and twice the second number.
9. If we assume that the sum of a number and its opposite is zero (example, $3 + {}^{-}3 = 0$), show that it must follow that ${}^{-}5 + 7 = 2$.
10. Show that the roots of $x^2 - 5x + 6 = 0$ are 2 and 3.

For Microteaching

One of the problems in using the three modes of explanation is the problem of switching from mode to mode in a flexible manner. In our discussion of modes, we concentrated on presenting an entire explanation in a single mode. In practice it is possible to move from mode to mode within a single explanation. For example, during the presentation of the symbolic proof of the irrationality of $\sqrt{2}$, we suggested that the teacher should include a discussion of relationships between fractions in an equivalence class. This move calls for a switch from the symbolic to the verbal mode within the same proof.

In other cases it may be wise to switch modes within an explanation because students indicate that they do not understand. For example, if students do not follow a verbal explanation of equivalence classes for fractions, it would be wise to switch to a graphic mode and locate equivalent fractions on a number line.

On the other hand, there are clearly times when one should *not* switch modes. Switching in the middle of a thought or idea can only add to confusion of the listener. An alternate explanation should help consolidate the understandings of basic ideas. Thus, it may be helpful to look at an explanation in terms of basic sections which are sequenced together. Each section should contain an important and complete idea. Modes can be switched between these sections, and entire sections may be repeated in a different mode. However, no switching should occur within sections.

The development of flexibility in modes of explanation requires both advance planning and practice. The teacher must study in advance to find alternate explanations for a topic. He must also practice keeping all these explanations in his head, being sensitive to the needs of students, and making quick decisions on the advisability of changing modes. These kinds of skills cannot come from "book learning" alone. The microteaching situation provides a good way to develop and practice these skills.

1. Choose a mathematics topic whose level of difficulty is appropriate for the students you will have to work with and which lends itself to several modes of explanation. You should look for a topic which should take no more than one or two minutes to explain in a single way — then plan on five or six minutes of microteaching time when other explanations are included. The three topics given in the *Application Module* may be used, or you may look at the section, *For Lesson Planning*, for other ideas.

2. Write a basic explanation of your topic. Divide your explanation into sections containing complete ideas (sections should seldom

be more than two or three sentences long). Plan an alternative explanation in a different mode for each section.

3. Present your topic to a class of five to ten students, recording your presentation on videotape. (An audio recording may be substituted if another observer can note and record your use of symbols, graphs, and models as you talk.) Try to be aware of individual student needs during your presentation by asking questions where appropriate.

4. View or listen to your presentation and identify the sections of your presentation that correspond to the sections of your original lesson plan. Play the tape again and decide the primary mode you used in each section. Study the variety and sequence of the modes you used. How might they be improved?

5. If possible, reteach your lesson to another group of students. Look at the tape of this lesson to identify changes you made.

For Related Research

The problem of matching lesson presentations to the abilities of different students is known as the problem of finding aptitude-treatment interactions. For an introductory discussion and overview of this research area, read Becker, Jerry P. "Research in Mathematics Education: The Role of Theory and of Aptitude-Treatment Interaction." *Journal for Research in Mathematics Education* 1 (Jan. 1970): 19-28.

The article which immediately follows Becker's gives details of a specific experiment designed to measure such interaction. Behr, Merlyn J. "Interactions Between 'Structure of Intellect' Factors and Two Methods of Presenting Concepts of Modular Arithmetic — A Summary Paper." *Journal of Research in Mathematics Education* 1 (Jan. 1970): 29-42.

For a more general approach to individual differences (including content areas outside mathematics), read Cronbach, Lee J. "How Can Instruction Be Adapted to Individual Differences?" *Learning and Individual Differences*. Edited by R. M. Gagné. Columbus, Ohio: Charles E. Merrill, 1967.

For a broader overview of aptitude-treatment interaction research, study the report by Cronbach, Lee J. and R. E. Snow. *Individual Differences in Learning Ability as a Function of Instructional Variables*. Stanford, Calif.: Stanford University Press, 1969.

We have indicated that the memory spans measured in the *Investigation Module* are surprisingly constant. For further evidence, read Miller, G. A. "The Magical Number Seven, Plus or Minus

Two: Some Limits on Our Capacity for Processing Information."
Psychological Review, 63 (1956): 81-97.

One of the natural ways to extend the experiment on immediate memory span is to see how the ability to remember a specific item (like a prime number, or a letter that is a vowel) depends upon the number of total items presented to the subject. An experiment like this is described by Lloyd, K. E., L. S. Reid, and J. B. Feallock. "Short-Term Retention as a Function of the Average Number of Items Presented." *Journal of Experimental Psychology*, 60 (1960): 201-207.

References

1. Bloom, Benjamin S., ed. *Taxonomy of Educational Objectives, The Classification of Educational Goals, Handbook I: Cognitive Domain.* New York: David McKay Co., 1956.
2. Guilford, J. P. *The Nature of Human Intelligence.* New York: McGraw-Hill Book Co., 1967.

Unit 2

Piaget's Analysis of Intelligence

Introduction

There is another and radically different way to view the components and structure of intelligence. This view holds that intelligence is structured not so much by the degree of different abilities as by the order in which certain abilities are developed or acquired. Most of the work in formulating this perspective has been done by the Swiss psychologist, Jean Piaget. The structure of the intellect as organized by Guilford views human abilities as relatively static. We can vary our teaching methods to fit these abilities, and in so doing, we may actually enhance latent abilities. Nevertheless, the final capacity in any area provides an upper boundary for our efforts.

Piaget's structure of the intellect is a much more dynamic theory. If a child does not have the ability to learn a particular concept, it is because this ability has not yet been acquired. The acquisition will be a function of both maturity and experience, and neither of these set an absolute upper boundary for the development of intellectual power. Although Piaget believes that in most cases acquisition can be accelerated by a teaching process, he has avoided the study of what such processes must be like. Instead, he has concentrated on finding the order in which intellectual abilities are acquired through maturation of the child and his normal experiences in the society.

The success of Piaget's studies depends first upon showing that children are *not* miniature adults. That is, they do not possess all the basic intellectual abilities of adults in diminished degrees, but, instead, they possess intellectual abilities that are actually quite different from the abilities of adults. Before studying the details of Piaget's work, one should first be convinced of these differences.

66

Conservation of
Numerousness and
Surface Area

(The success of this investigation depends upon locating young children between the ages of four and fourteen to act as subjects. We will be primarily interested in showing the existence of striking differences between children of different ages, rather than establishing precise ages at which changes occur. Therefore, finding a few children whose ages are uniformly distributed in the four to fourteen range is more important than repeating trials with many children of approximately the same age. For this reason the investigation may be assigned to the entire class as a group project rather than to each individual.)

The experiment in this investigation consists of presenting the subjects (children) with two simple tasks: to determine whether there are more objects according to how they are piled together or separated, and to determine whether or not more paint would be required to cover the outside of some stacked arrangements than other arrangements. We will use an interview technique in which the interviewer asks these questions of the children in an informal conversational manner. If several interviewers are to pool their results, they should agree on the way the questions will be worded. Once the "standard question" is asked, however, the interviewer may ask the question in other words to be sure that the initial response was not due to a misunderstanding of vocabulary. For objects we will use sugar cubes which are extremely regular and relatively inexpensive.

The interviewer should be seated at a table with the subject seated on the opposite side. No other children should be in the immediate area at the time of the interview. Two sheets of paper should be placed on the table. Label one sheet "A " and the other sheet "B" so that the subject can clearly read the labels. Twelve sugar cubes are placed upon each sheet of paper according to the arrangement key.

Here is a suggested set of questions that interviewers should ask. The questions may be modified, as appropriate, if all interviewers make the same changes.

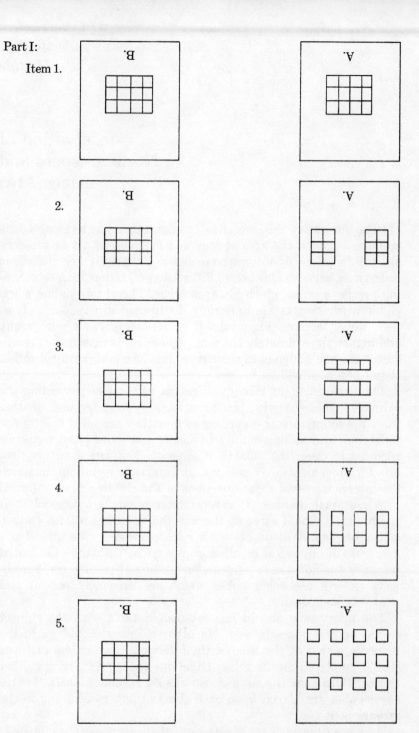

Figure 14. Test for Numerousness and Surface Area.

Part II:

6.

B.

A.

7.

B.

A.

8.

B.

A.

9.

B.

A.

10.

B.

A.

When the subject is seated, the interviewer introduces himself, and says, "I want to find out some of your ideas about blocks and cubes like these sugar cubes." (Show some of the cubes.) "Here are two pieces of paper, A and B." (Place the sheets of paper on the table with the labels facing the subject.) "Would you point to the piece of paper labeled B?" (This checks that young subjects can read the labels, but may be omitted for older subjects.)

"Now here are the sugar cubes." (The cubes are placed in two 3 x 4 blocks as shown in item 1 of the arrangement key.) "Does one piece of paper have more sugar than the other, or do they have the same amount of sugar?" (If the response only indicates that they are different, ask which one has more.)

"Suppose that you had a small can of paint and were to paint the outsides of the piles of cubes. Would you need more paint for the sugar on one piece of paper than for the sugar on the other piece of paper, or would you need the same amount of paint for each?"

After recording the subject's responses for each item, the inter-viewer moves the sugar on paper A so that two piles are formed on the paper like the configuration shown in item 2. The two questions may be shortened to:

"Now is there more sugar on paper A or paper B, or the same amount of sugar on both?"

"Would you need more paint to cover the outsides of the piles on paper A or on paper B, or the same amount for both?"

During the experiment the interviewer should mark down the response of the subject to both questions of each item by writing down an "A," "B," or "S" according to whether the subject's answer indicated more sugar or paint for paper A or B or the same amount for both. Immediately after the session the interviewer should note the rapidity or hesitancy shown by the subject in answering items, the nature of the questions asked by the subject (or lack of such questions), and any unusual procedures or actions shown by the subject during the interview.

When all subjects have been interviewed a table should be made listing them in order of increasing age and indicating whether their response to each item was right or wrong. The following column headings are suggested: Age—Years: Months; Sex; Quantity Questions—1, 2, 3, 4, 5, 6, 7, 8, 9, 10; Area Questions—1, 2, 3, 4, 5, 6, 7, 8, 9, 10.

Analyzing the Data

The following table was made when this experiment was run at Stanford University in the spring of 1969 using children of Stanford students as subjects. You should not expect your results to match these in every detail since there may be considerable differences in the previous experiences of your subjects and these children. None of these children had had specific training in conservation tasks, however, so you should be able to identify similarities and trends between this data and the data you have just collected.

Age Yr:M	Sex	Quantity Questions 1 2 3 4 5 6 7 8 9 10	Area Questions 1 2 3 4 5 6 7 8 9 10
5:1	F	x w w w w * * * * *	x w w x w * * * * *
5:10	M	x w w w w * * * * *	x w w w w * * * * *
7:0	M	x x x x x * * * * *	x w w w w * * * * *
8:7	F	x x x x x * * * * *	x w w w w * * * * *
10:0	M	x x x x x x x x x x	x x x x x x x x x x *
10:4	F	x x x x x * * * * *	x w w w w * * * * *
11:0	F	x x x x x x x x x x	x x x x x x x x x x
11:2	F	x x x x x x x x x x	x x x x x x x x x w
13:2	F	x x x x x x x x x x	x x x x x x x x x x
16:6	F	x x x x x x x x x x	x x x x x x x x x x

x = correct response
w = wrong response
* = the subject was not tested on this item

Table 6. Stanford University Test Results

In this data three general levels can be observed. In the first of these, the subject believes that the number of objects is somehow related to their spatial arrangement. If the objects are spread out over a greater area, the subject believes that there are more of them. Questions relating to the surface areas of various piles are answered incorrectly or irregularly. Most subjects give rapid responses to the questions with little time for reflection if the basic question is understood. Rapid guessing appears to be the major technique employed. This level lasts until approximately age six.

At that time a change occurs. In the second level the subject seems to understand well that the number of objects displayed is

independent of their spatial orientation. This concept is known as
conservation of numerousness. The subject can physically dissect
the sugar cube structure and rearrange its component pieces, yet
realize that the total number of blocks is unchanged. This ability
opens the way to *concrete operations* with objects. Since spatial
orientation does not affect numerousness, the child can form equivalent structures that are easier to analyze. He can also use symbols
to record his operations. Our data indicates, however, that children
in this level may overextend the general conservation concept. Their
responses to questions related to the surface area of the piles show
that they also believe there is no relationship between the surface
area of the piles and the spatial orientation of cubes making up
those piles. Since the number of cubes is the same in each case,
they believe that the paint required would also be the same.

At about age ten another change occurs. This change can usually
be seen not only in the subject's responses but also in the way he
responds. He listens intently to the interviewer's questions and is
not so quick to answer. The questions relating to surface area in
particular are apt to provoke periods of quiet reflection. The subject asks for clarification, and that he is weighing a variety of possibilities is often clear. When answers are given, the questions relating
to surface area are answered consistently and correctly. This third
level seems to be characterized by an ability to think abstractly
about the problem with the situation varied mentally in order to
judge the various outcomes. The child has progressed from an
ability to analyze by actually moving the physical pieces (concrete
operations) to an ability to analyze the consequences if the pieces
are changed mentally. This level is known as *formal operations*.

These three levels form the skeletal outline of Piaget's work. The
complete outline must encompass much more. In particular, we
must be sure that our "levels" are not merely functions of the particular content ideas we were investigating — numerousness and
surface area. Would we find a similar development of abilities where
other mathematical concepts are concerned? What about matching
by one-to-one correspondence, for example? Can this matching
be done concretely before it can be done mentally or formally by
physically pairing up objects?

The Child's
Formation of Basic
Mathematical
Concepts—
Psychological and
Mathematical
Structure

In a now famous article in *Scientific American*, Piaget traced the development of mathematical ideas in young children as revealed by his experiments. He maintains:

It is a great mistake to suppose that a child acquires the notion of number and other mathematical concepts just from teaching. On the contrary, to a remarkable degree he develops them himself, independently and spontaneously. When adults try to impose mathematical concepts on a child prematurely, his learning is merely verbal; true understanding of them comes only with his mental growth. (8:74)

A good example of verbal learning is the way young children learn to count from their parents. These children can correctly count ten objects laid in a row, but cannot count the same objects correctly when they are piled up or arranged in a complex pattern. The child knows the names of the numbers, but he does not understand one of the most fundamental ideas of number — that the number of objects does not depend upon the spatial arrangement of those objects. No matter how the objects may be moved or shuffled, their number remains the same. Piaget calls this fundamental property *conservation of numerousness*.

In contrast, a six- or seven-year-old can show that he has spontaneously developed this concept even though he may not be able to count in a formal way. Given a pile of red chips and a pile of blue chips and asked to find out which pile contains more, he will proceed to match one red chip to one blue chip until one of the piles is exhausted. This one-to-one correspondence procedure is essential to counting, yet it does not require a knowledge of the names of numbers. Once the child develops the correspondence procedure he can show that two groups of objects can remain equal in number no matter what shape they take.

If we modify the experiment slightly, we can investigate the development of conservation of numerousness in young children. We may lay down a row of eight red chips, equally spaced about an inch apart, and ask the children to take from a box of blue chips just as many chips as we have placed on the table. Piaget finds:

Their reactions will depend on age, and we can distinguish three stages of development. A child of five or younger, on the average, will lay out blue chips to make a row exactly as long as the red row, but he will put the blue chips close together instead of spacing them. He believes the number is the same if the length of the row is the same. At the age of six, on the average, children arrive at the second stage; these children will lay a blue chip opposite each red chip and obtain the correct number. But they have not necessarily acquired the concept of number itself. If we spread the red chips, spacing out the row more loosely, the six-year-olds will think that the longer row now has more chips, though we have not changed the number. At the age of six and a half to seven, on the average, children achieve the third stage: they know that, though we close up or space out one row of chips, the number is still the same as in the other. (8:74)

This phenomenon does not depend upon the nature of the experimental materials or design. We may give the child two identical containers and ask him to fill them with black and white beads one at a time by simultaneously putting a black bead in with his left hand and a white bead in with his right hand. When the containers are about full we ask him how they compare. He is certain that both containers have the same number of beads. But, if the black beads are poured into a third container which differs in size and shape and the question repeated, we see a difference in understanding according to the age of the child. The youngest children think that the number of black beads has changed. If the level of beads is higher in the third container, they think there are more; if lower, they think there are less. But by the age of seven, most of these children will know that the transformation has not changed the number of beads.

In short, children must grasp the principle of conservation of quantity before they can develop the concept of number. Now conservation of quantity of course is not in itself a numerical notion; rather it is a logical concept. Thus these experiments in child psychology throw some light on the epistemology of the number concept — a subject which has been examined by many mathematicians and logicians. (8:75)

In general, a mathematician's analysis of the concept of number, based on the logical foundations of mathematics, differs from Piaget's psychological analysis. For example, Poincare and Brouwer held that the number concept is a product of primitive intuition which *precedes* logical notions. Piaget's experiments would deny this theory and suggest that the development of number depends upon the child's logical considerations. Piaget's analysis would place him closer to Bertrand Russell who held that number is a purely logical concept. Russell distinguished two basic concepts underlying number—cardinal numbers and ordinal numbers. The idea of cardinal number comes from the logical notion of category or set, while the idea of ordinal number comes from the logical relationships of order. But Piaget does not find that children make a distinction between cardinal and ordinal numbers. In addition, he argues that the concept of cardinal number presupposes an order relationship.

For instance, a child can build a one-to-one correspondence only if he neither forgets any of the elements nor uses the same one twice. The only way of distinguishing one unit from another is to consider it either before or after the other in time or in space, that is, in the order of enumeration. (8:75)

This argument does point out a basic difference between the logical analysis by a mathematician and the psychological analysis by Piaget. For Piaget, the equivalence of sets comes as a result of a one-to-one correspondence procedure by the subject. He has focused upon the *action* used by the subject where the concept is involved. Russell, on the other hand, interprets a number as a set of equivalent sets. The physical process by which the component sets were judged to be equivalent is of no consequence to him if the *idea* of one-to-one correspondence is clear.

Relationships between logical and psychological analyses of mathematical ideas are also apparent if we study the child's discovery of spatial relationships or geometric concepts. The historical development of geometry began with the Euclidean system, was extended in the seventeenth century with the consideration of perspective (projective geometry), and continued in the nineteenth century with topology which describes spatial relationships in terms of open and closed curves, interiors, exteriors, and separation.

Piaget finds that the child's development of geometry reverses this historical order. The child's first geometric distinctions are topological. A three-year-old can distinguish between open and closed figures. He can reproduce figures showing a small circle inside or outside of a larger circle. But he cannot distinguish between a triangle and a rectangle. When asked to reproduce these, he simply responds by drawing a lopsided curve. He does not develop ideas of Euclidean and projective geometry until after he has mastered topological relationships.

Piaget's tests for projective constructions are as ingenious as his tests for number concepts. One involves the use of "fence posts," little sticks stuck in bases of modeling clay. Two posts are placed about fifteen inches apart, and the child is asked to place other posts in a straight line between them. Children under the age of four proceed to place one post next to another, making a wavy line. They join the individual posts on the basis of proximity instead of projecting a line from the original end posts. At the next stage, after the age of four, the child may be able to form a straight line between the two end posts if he has some other straight line to guide him. For example, he will succeed at the task if the two end posts form a straight line which is parallel to the edge of the table, or to the wall. But if the end posts are diagonally across the table, he may begin by placing posts parallel to the table's edge and then, suddenly shifting direction, form a curve to reach the second end post. Not until the child is about seven, on the average, can he check the straightness of a fence by sighting along it. Not until this time has the child understood that the projective relationship depends upon the angle of vision or the point of view.

The inability of children to recognize different points of view is illustrated by an experiment in which the experimenter sits at a table opposite the child and places a cardboard cutout representing a range of mountains between them. The two see the mountains from opposite perspectives, of course. The child is asked to select from several drawings the one which pictures his view of the mountain and the one which pictures the experimenter's view of the mountains. The youngest children can pick out only their own view of the mountains; they believe that the experimenter must also be viewing the mountains just as they are. If the child changes places with the experimenter, he can correctly pick out the picture which corresponds with his new view, but believes that this new view is the only correct one. He is not able to reconstruct the point of view which he previously held. The ability to coordinate different perspectives is not developed by the child until the ages of nine or ten. This example of egocentricity is characteristic of the primitive reasoning of children which keeps them from understanding that there may be points of view other than their own.

Where is the conservation principle in all of this? Piaget notes that when many children form the concept of projective space they line up the fence posts not only by sighting but also by lining up his hands parallel to each other. The hands are being used to preserve the direction; Piaget calls this concept *conservation of direction.*

Young children differentiate between conservation of length and conservation of distance. Distance concepts can be investigated by placing a block or cardboard wall between two toy trees standing on a table top. The child is asked whether or not the trees are still the same distance apart, now that the wall separates them. The youngest children think that the distance has changed; they cannot relate the two parts to the original. Children of five or six believe that the distance between the trees has been reduced. Their justification of this belief is that the width of the wall does not count as distance — there is a difference in value between empty space and filled-up space. Not until they are near the age of seven do children realize that intervening objects do not change the distance. No matter how they are tested, Piaget finds that

...children do not appreciate the principle of conservation of length or surface until, somewhere around the age of seven, they discover the reversibility that shows the original quantity has remained the same (e.g., the realignment of equal-length blocks, the removal of the wall, and so on). Thus the discovery of logical relationships is a prerequisite to the construction of geometrical concepts, as it is in the formation of the concept of number. (8:76-77)

Concepts of distance and of length are related, of course, by processes of measurement. Piaget finds that concepts related to measurement also go through a fascinating evolution. A child is shown a tower of blocks on a table, and is asked to build a second tower of the same height on another table whose top is lower or higher than the top of the first table. The blocks available for building the second tower are of a different size than those used to build the first tower. Measuring tools are available for the child's use. Very young children will build the second tower up until both tops appear to be at the same level, without considering the different heights of the table tops. Later he may lay a long rod across the tops of the two towers to check them. Still later he will notice that the two table tops are not at the same height and begin to look around for a measuring standard. The first one that usually comes to mind is his own body. He tries to "carry" the height of one tower over to the other by spreading his hands apart and moving them to the other tower. When he finally notices that this method is not very reliable, he will measure the tower against reference points on his body. This reaction is fairly typical of six-year-olds. Eventually, he will construct a third movable tower which can be matched against both his construction and the original model. This presupposes a process of logical reasoning. If the height of the model tower is A, the second tower C, and the movable tower B, he has reasoned that B = C and B = A, therefore, C = A.

Not until sometime later will the child realize that he can use a shorter rod to measure the tower. To measure, the child must see the whole as a number of parts added together, and be able to substitute one part for others. Piaget concludes,

... measurement is a synthesis of division into parts and of substitution, just as number is a synthesis of the inclusion of categories and of serial order. But measurement develops later than the number concept, because it is more difficult to divide a continuous whole into interchangeable units than to enumerate elements which are already separate. (8:78)

Not surprisingly, the child finds measurement in two dimensions even more difficult. Given a sheet of paper with a dot on it, and asked to put a dot in the same position on a similar sheet of blank paper, he is either satisfied with a visual approximation, or with a single measurement of the distance of the dot from the bottom of the paper. He is surprised that this single measurement does not give him the correct position. He may then measure the distance of the dot from a corner of the paper, trying to keep the same angle or slant to his ruler at all times. Not until he is eight or nine does he realize that the measurement must be broken up into two operations. The measurement of a plane with a Cartesian coordinate system depends upon the horizontality or verticality of physical objects. But even these basic concepts are slowly developed in children. Using a jar half-filled with colored water, Piaget asks children to predict what level the water will take when the jar is tipped in different directions. Not until he is nine, on the average, does the child grasp the idea of horizontality, and correctly predict that the surface of the water will always remain horizontal. With this attainment, the child completes his conception of how to represent space.

Piaget's Intellectual Structure

Piaget likes to refer to the method he uses in his experiments as clinical. The method is perhaps best described as a combination of observation and interview. He examines children in an unstructured situation in which the experimenter follows the lead of the child's responses and makes a verbatim record of the interview. These records are analyzed and categorized according to content and observed behavior. The results are usually reported by giving examples of the categorized verbatim records, and Piaget follows these reports with an explanation of the process used by the child. This analysis and explanation is usually very detailed so that Piaget's writings tend to become very lengthy. He seldom gives statistical information to clarify how many children at a specified age were placed in each category, or how many children were tested. This method has not only brought criticism to his work but also makes the task of summarizing his results very difficult.

It is possible to give a rough definition of Piaget's principal scientific concerns in a single sentence. He is primarily interested in the theoretical and experimental investigation of the qualitative development of intellectual structures. In Piaget's system, the vast array of changing structures that occur in the course of development are divided up into stages whose qualitative similarities and differences serve as conceptual landmarks in trying to grasp the process. Piaget believes that within the intellectual development of any individual these stages emerge in an unchanging order or sequence. Although the sequence itself is taken as invariant, the age at which any particular stage appears may vary considerably from individual to individual and from society to society.

The sequence of Piaget's stages and the observational experiments which led to this sequence are spread out over more than twenty-five books and one hundred fifty articles. The most complete summary and analysis of this work is contained in John Flavell's book, *The Developmental Psychology of Jean Piaget.* (2) Piaget himself has also prepared several summaries. Perhaps the best of these was prepared for a special conference at Cornell University and appears in the *Journal of Research in Science Teaching.* (7) In the following overview, the term "period" is used to designate the major developmental sections, and the term "stage" is used to denote the smaller subdivisions within the periods.

1. *The Period of Sensory-Motor Intelligence*
   (0-2 years)

During this important first period, the infant moves from a neonatal, reflex level where there is no differentiation between self and world to a relatively coherent organization of sensory-motor actions which recognize the immediate environment. The organization is an entirely practical one, however, in the sense that it involves simple perceptual and motor adjustments to things rather than symbolic manipulations of them. There are six major stages in this period. Stages 1 and 2 constitute the time span in which the infant is locked up in his own egocentrism. He is confined to surveying what must be an orderless array of stimulation without really being able to act on things and observe, in even the most limited way, how these actions interact with the things they contact. In Stage 3 he begins to move out into the unknown medium which surrounds him, thanks primarily to the growing, all-important ability to direct his hand movements. In doing so, he begins to be able to perceive simple connections between the two realms of self and the outside world. Of course, this perception is still very egocentric in that the child apprehends the two only as an undifferentiated whole. Nevertheless, this movement outward is full of cognitive possibilities which the infant will gradually exploit. Stage 4 is a transitional stage of decisive progress in substituting object-object relations for subject-object ones. The child begins to see things relate to other things, still in the context of his own action but increasingly independent of it. With the lengthening of time between original intention and final objective goal, the gap between self and world widens and the action-object connections begin to break apart. Stage 5 is really the culmination, as far as sensory-motor development in the strict sense is concerned. Objects are now really detached and independent entities which can be imitated, inserted into play schemes, and related spatially, temporally, and causally. The self also begins to be treated like other objects, as something with its own texture and resistance, its own locomotion relative to fixed object positions in space, and so on. Finally, the child of Stage 6 crowns these achievements with added finesse and skill and enriches them through the powerful tool of an emerging capability for symbolization. In doing all these things, the child has passed into a new era in which this symbolic capacity will become the important instrument of cognition.

2. *The Period of Preparation for and Organization of Concrete Operations (2-11 years)*

This period begins with the first crude symbolizations late in the sensory-motor period and concludes with the beginnings of formal thought in early adolescence. There are two important subperiods. The first, that of preoperational representations (2-7 years), is that period in early childhood in which the individual makes his first relatively unorganized and fumbling attempts to come to grips with the new and strange world of symbols. Piaget sometimes distinguishes three stages in this first subperiod: the beginnings of representational thought (2-4 years), simple representations or intuitions (4-5½ years), and articulated representations or intuitions (5½-7 years). The operations of classification, ordering, the construction of the idea of number, spatial and temporal operations, and all the fundamental operations of elementary logic of classes and relations, of elementary mathematics and geometry, and even of elementary physics appear in this subperiod. The operations are concrete, however, because they operate on objects and not yet on verbally expressed hypotheses. The child's conceptual organization slowly takes on stability and coherence by the forming of a series of cognitive structures. Piaget has constructed models of these structures which he calls *groupings*. These *groupings* are combinations of operations which satisfy the requirements for a mathematical group or a mathematical lattice when considered individually. However, when the operations are combined into Piaget's *groupings*, the resulting structure is neither a mathematical group nor a mathematical lattice although they retain some properties of both. The most striking mathematical property is that of reversibility. For Piaget, reversibility is *the* core property of the cognition system. In this subperiod the child first begins to appear rational and well-organized in his adaptations. He appears to have a fairly stable and orderly conceptual framework which he systematically brings to bear on the world of objects around him.

3. The Period of Formal Operations (11-15 years)

During this period a new and final reorganization takes place, with new cognitive structures appearing which are isomorphic to the groups and lattices of logical algebra. In brief, the adolescent can deal effectively not only with the reality before him (as does the child in the preceding subperiod) but also the world of pure possibility, the world of abstract propositional statements, the world "as if." The child can reason on hypotheses as well as objects. He constructs new operations, operations of propositional logic rather than simply the operations of classes, relations, and numbers. The taking into account of all possible combinations, testing and rejecting each in turn, is typical of this period. This kind of cognition, for which Piaget finds considerable evidence in his adolescent subjects, is adult thought in the sense that these are the structures within which adults operate when they are at their cognitive best, that is when they are thinking logically and abstractly.

Each of the stages summarized is also characterized as containing an initial period of preparation and a final period of achievement. The preparatory phase, which is denoted by flux and instability, eventually gives way as the structures in question form a tightly knit, organized, and stable whole. The structures which have defined earlier stages in development are also integrated into these following stages. Finally, when these structural properties

attain stability or an equilibrium state they characteristically
show a high degree of interdependence. This unified and organized
character of structures would seem to make possible the definition
of this totality which they form. Piaget's goal has been to find those
totalities or structural wholes which correctly identify the essences
of organized intelligence at their various levels.

Piaget also recognizes another aspect of stage development which
he calls *decalage* or temporal displacement. *Horizontal decalage*
means that once an individual attains a particular cognitive structure, he may not be able to perform within that structure for all
tasks. For example, the recognition of the conservation of mass (or
quantity of matter) and the conservation of weight would imply
the same certain cognitive structure. Yet the conservation of mass
is typically achieved by children a year or two earlier than the conservation of weight. On the other hand, repetition may occur at
distinctly different levels of functioning, and this concept is called
vertical decalage. The young child develops a precise behavioral
map of his immediate surroundings, yet it will be several years before he can represent the terrain and its relationships symbolically
by drawing a simple map. Despite this time difference, there are
clear similarities in the reality content and in the cognitive organization necessary for both tasks. The implication of *vertical decalage*
is that there is a hidden uniformity within the apparent differences
between one stage and another. There seems to be little in common
between the groping, stumbling, exploring walks of a toddler and
a map-making project in which a fifth grader participates. Yet there
are structural similarities buried in the obvious differences, and it
is this recurrence which defines vertical decalage.

Piaget does see some constancy in intelligence. There are broad
characteristics of intelligent activity which hold true for all ages,
and which virtually define what is meant by intelligent behavior.
Piaget calls these broad characteristics *intelligence function.* He
sees that the fundamental properties of intellectual functioning are
always and everywhere the same, despite the wide varieties of cognitive structures which are created by this functioning. Piaget calls
the fundamental properties of functioning *functional invariants.*
There are two of these properties which are considered basic: organization and adaptation. In addition, Piaget divides adaptation
into two subproperties, assimilation and accommodation. Assimilation occurs when new information is taken into already existing
patterns and structures.

Piaget sees assimilation at work, for example, whenever a situation
evokes a particular pattern of behavior because it resembles situations
that have evoked it in the past, whenever something new is perceived

or conceived in terms of something familiar, whenever anything is invested with value or emotional importance. Accommodation, on the other hand, means the addition of new activities to an organism's repertoire or the modification of old activities in response to the impact of environmental events. (1:2)

At the other end of the scale is intellectual content. By "the content of developing intelligence," Piaget means raw uninterpreted behavioral data. Thus, the child's actual answers in Piaget's clinical interviews or the sensory-motor actions of the child in solving a physical problem are a part of the content of intelligence.

Interposed between function and content are the cognitive structures. These structures in Piaget's system are the organizational properties of intelligence, which are created through functioning. Since these organizations determine the nature of behavior contents, Piaget reasons that they are inferable from such contents. They are sometimes referred to as mediators interposed between the invariant functions and the variegated contents which appear through behavior. Like content (and unlike function) these cognitive structures do change with age, forming the periods and stages we have just summarized.

The cognitive structures are fundamentally similar to mathematical structures. *Transformation* is the key idea in all cognitive structures. What happens when an object is transformed to another state or when an idea is applied to a different situation is, for Piaget, the basic question used by the child to generate intellectual development. How similar this structure is to mathematics, whose fundamental concept, *function*, is illuminated by looking at *mappings* from one set to another! Piaget believes that the most powerful tool for understanding transformations is the idea of *reversibility* of transformations. Testing transformations for reversibility allows the child to examine variables for constancy and *equivalence*. But these elements are precisely the tools used for studying mathematical functions and mathematical structures.

The layman commonly believes that mathematics is basically foreign — foreign to his everyday experiences and foreign in its language of expression. But Piaget's work turns the tables on our intuition. The basic structures of mathematics are analogous to the basic cognitive structures used in intellectual functioning by every human being. Mathematics is not just the formalized expression of thought processes used by a small group of unusual individuals employed in departments of mathematics. Mathematics is the formalization of basic structures which underlie the thought processes used by every child and every adult. The difference in intelligence between children and adults is that some of these structures are in the process of developing for children. This development places the classroom teacher in a dilemma. On the one hand, mathematics is "natural" for children and should stand at the center of our school curriculum. On the other, the cognitive structures necessary for learning much mathematics may not be present at all in the early elementary school years and will be present only in differing stages in the junior high school and early senior high school years. In fact, Karplus has recently done work which suggests that many adults never fully develop the ability for abstract symbolic reasoning which characterizes the ultimate development in the final period of formal operations. (6:400) Must we then teach what cannot be taught? The resolution of this dilemma lies in Piaget's concept of knowledge itself. That concept of knowledge and its implications for classroom teaching are discussed in the following Application Module.

Discovery Teaching
and Mathematics
Laboratories

Piaget's life work has been the study of knowledge and how it develops in the human organism. He classifies himself as a *genetic epistemologist*. To consider the work of Piaget as a learning theory is technically erroneous. But it is often hard to make a complete distinction between how intelligence develops and how things are learned. If we assume that learning should parallel the development of intelligence, we can rather easily deduce from Piaget's work some of the principles that such learning should follow.

We should first take into account the child's intellectual capacity as implied by the current stage in his development. For example, we commonly illustrate the addition operation by combining groups. To show the addition of $3 + 5$, we place three stones in one pile, five stones in another pile, and add them by sweeping them together into a single pile. We count the stones in this final pile and triumphantly announce that this shows that $3 + 5$ must equal 8. But have we shown any such operation? Suppose the child has not yet achieved the principle of conservation of numerousness. Then he believes that the number of stones depends upon their spacing; usually that there are more stones in the two original piles than in the final one. What we have shown such a child is that it is somehow possible to squeeze three stones and five stones into eight stones. We have certainly not (for his intellectual capacity) shown that $3 + 5$ and 8 are two names for the same number. What alternatives are left for this child? He must either rotely memorize a set of addition facts or develop a process or algorithm for arriving at these relatively meaningless facts. If the teacher expects more, he runs the risk of producing frustration and number anxiety which could be the foundation of the all-too-common "I can't do arithmetic" reaction. The teacher should always consider whether or not the student is at a developmental stage which would allow him to meet expectations. In the course of his studies, Piaget has created many cognitive tasks to test aspects of the intelligence of children. Researchers are presently developing and standardizing such tests which promise to be of great use in grade or group placement or in identifying students who need remedial training programs. (3:115)

But, beyond accounting for the child's intellectual stages, we should also make our teaching methods consonant with the basic nature of the knowledge we are teaching. What is that nature? For Piaget, to know an object or to know an idea is to transform it, to manipulate it, to change it. It is not just learning but knowledge itself that is *active*. We cannot know by passively receiving. We know only by acting and testing. Have you studiously taken notes in a lecture, only to find out later that what you needed for knowledge was not the steps written down, but the way the lecturer manipulated ideas *between* steps? Or have you discovered that seeing a problem solved in class means nothing until you have struggled with the same problem yourself? Of course, it is possible to learn from a lecture or from seeing a great problem solver in action, but in these cases, the listener follows the classroom action by mentally manipulating, transforming, and testing ideas as they are presented. If knowledge itself is active, certainly learning can be no less active. The key to learning must be experience.

But what kind of experiences are best teachers? One can identify three general types of classroom teaching experiences: explanatory, inductive, and incidental. We have already discussed explanatory teaching in terms of modes of explanation. Methods which use inductive and incidental techniques are commonly called *teaching by discovery*. The distinction between explanatory teaching and discovery teaching is far from a sharp one. A learner can be active and can make discoveries while listening to a lecture. Yet lecturing is not usually classified as a discovery teaching technique because most of the burden of manipulation and transformation of ideas so necessary for discovery rests upon the listener during a lecture. The formal lecturer who presents his ideas in an organized and final form, who does not entertain questions or suggestions from his audience, does practically nothing to encourage mental activity on the part of his listeners.

In contrast, the lecturer who encourages discussion in his class permits learners to engage in active consideration of ideas. Best of all is the teacher who sees himself not as a lecturer but as a traffic monitor at the center of a great intersection. The intersection is formed by the students in his class; the traffic is traffic of ideas which are being actively tried, manipulated, changed, and tested. The role of the traffic monitor, like the role of a traffic cop, is only to keep things moving. He encourages potentially productive ideas, discourages thinking that can only lead into intellectual deadends, and emphasizes conflicting ideas which must be tested and transformed before further progress can be made. Such a teacher does *not* need to have the subject matter thought out in advance (nor have a lesson extensively planned). On the contrary, the teacher who acts as a traffic monitor in ideas must know far *more* subject matter than the lecturer, and he must have planned for many different contingencies that may be triggered by students' ideas. Such a teacher is not really a discussion leader, for he is not so much ahead and leading the discussion as he is inside the discussion monitoring the flow of ideas. The Piagetian teacher's concern is ultimately one of providing *environments for learning* which allow learners to manipulate, either concretely or abstractly, and, thus, to know. In a sense, he is a learning chemist who creates the proper mix in a test tube or classroom and then removes himself as much as possible to watch the reaction and to monitor its results.

Encouraging Learning by Discovery

Unfortunately, the admonition to "stand back and let discovery happen" simplifies the matter unfairly. Many classroom hours pass in which mathematical ideas are not spontaneously produced by students and ideational conflicts do not arise that require testing and resolution. Active learning must be encouraged by the teacher in such classes. Making use of inductive teaching methods is one way of encouraging active learning. The key to inductive teaching is *sequence*. The teacher sequences the details of the lesson in small steps so that all the data needed for a generalization are made available to the students. By skillful use of questions or problem sets, he encourages the learners to take an active role in relating ideas.

Examples of discovery teaching by induction are plentiful. Perhaps the oldest example is contained in Plato's *Dialogues* in which a demonstration lesson taught by Socrates is recorded (5:361-366). The problem posed for the lesson is that of finding a square whose area is double that of a given square. The given square is two units on each side. Socrates asks the boy how the area of the given square compares with the area of a square whose sides measure only one unit. He also asks the boy to double the sides of the original square and count the resulting area. From these activities the boy concludes that doubling the side of a square gives an area four times the original.

Having discovered that a square four units on a side gives more than the doubled area being sought, Socrates then asks if the side of the desired square should be somewhere between the two units of the original square and the four units of the last square considered. The boy agrees to this, and Socrates suggests drawing a square whose sides are three units. This is done, and the boy determines that the area of this square is nine square units — still more than double the area of the original square.

At this point Socrates comments that all he has done is convince the boy that he does not know how to construct a square whose area is double that of the original. Nevertheless, the boy is better off than he was originally because he is now perplexed about the problem and is motivated to inquire about it. This is an important property of a good induction sequence. It should reveal the important difficulties and aspects in the problem, and in so doing motivate the student to find a solution. Induction sequences which omit these steps usually appear to be little more than drill exercises.

Finally, Socrates draws the diagonal of the original square and asks how it divides the area. By forming the square of the diagonal, he reaches his desired conclusion. Socrates has taught the lesson using only his questions and the boy's responses. This questioning technique is known as the Socratic method. The elements of induction in this ancient example are clear. The lesson begins with something already known by the student, and proceeds in small steps to emphasize the important variables which must be related to form the desired conclusion or discovery.

Math 3   Name_____

Today we will look for a shortcut to use when multiplying
by 5. Do all of the following problems.

1. Find the answers:

    (a) 14   (b) 25   (c) 72   (d) 117   (e) 136
        x5   x5   x5   x5   x5

2. Find the answers:

    (a) 14   (b) 25   (c) 72   (d) 117   (e) 136
        x10   x10   x10   x10   x10

3. (a) Subtract the answer for 1(a) from the answer for 2(a): ____
    (b) Subtract the answer for 1(b) from the answer for 2(b): ____
    (c) Subtract the answer for 1(c) from the answer for 2(c): ____
    (d) Subtract the answer for 1(d) from the answer for 2(d): ____
    (e) Subtract the answer for 1(e) from the answer for 2(e): ____

4. Do the following problems without multiplying by 5. (If you
    need help, raise your hand.)

    (a) 18 x 5 = ____   (d) 99 x 5 = ____

    (b) 32 x 5 = ____   (e) 103 x 5 = ____

    (c) 81 x 5 = ____   (f) 159 x 5 = ____

5. Did you subtract as part of doing section 4? Could you do
    section 4 without multiplying by 5 and without subtracting?
    (If you need help, raise your hand.) Do the following problems
    as quickly as you can:

    (a) 13 x 5 = ____   (h) 93 x 5 = ____

    (b) 21 x 5 = ____   (i) 102 x 5 = ____

    (c) 38 x 5 = ____   (j) 112 x 5 = ____

    (d) 46 x 5 = ____   (k) 127 x 5 = ____

    (e) 63 x 5 = ____   (l) 132 x 5 = ____

    (f) 75 x 5 = ____   (m) 145 x 5 = ____

    (g) 87 x 5 = ____   (n)1284 x 5 = ____

Figure 15. Mathematics Problem Sheet.

Often we can induce discovery in mathematics by sets of carefully sequenced problems. Consider Figure 15.

Of course, we cannot always be sure that problem sheets like this one will induce discovery by students. What are some of the important considerations in sequencing problem sets?

1. The timing for giving the problem set is crucial. Induction sequences build on knowledge and skills which the student already possesses, and he extends them into new areas. This sample problem set depends upon the ability of students to multiply by 5 and by 10, to subtract accurately, and to divide by 2. What would happen if this problem set were given to children who could not accurately multiply by 5? Mistakes in section 1 would make it absolutely impossible to discover a pattern in section 3. The teacher who uses discovery techniques must know the capabilities of his students very well. If he begins with problems beyond the abilities of his students, his induction sequence will build nothing but confusion and frustration.

2. Discovery requires that the end result can be achieved in more than one way. The discovery sought in this problem sheet is that multiplication by 10 followed by division by 2 is an easy way to accomplish multiplication by 5. Yet if we simply instruct the student to "first multiply by ten, then divide by two and see what happens," we will have removed almost all of the discovery. Clearly what is needed is an alternate way to arrive at this process. In our example, subtraction provides that alternate to division. As a result, the student is required to do more than just follow instructions. He must make connections and supply relations himself. These processes lie at the heart of discovery.

This requirement for alternate methods implies that some areas of mathematics are *not* suitable for discovery! In particular, to expect students to discover arbitrary definitions, postulates, or assumptions is futile. The fact that these things are arbitrary conventions means that they cannot be uniquely determined by alternate approaches. However, even if one cannot discover definitions and conventions, discovering why certain definitions are plausible or why certain conventions are reasonable to make is possible. Such plausibility is easy to develop by looking at the consequences of alternate definitions and conventions. This kind of exploration is best done as a class discussion, however.

3. A discovery sequence must include enough cases for complete generalization. In our example problem sheet, both even and odd multiplicands have been included. Often in haste, only even numbers will be included in such a problem sheet. As a result, observant students might conclude incorrectly that the discovered method worked only for even numbers.

Sometimes the difficulty may occur in reverse. If the teacher wants students to discover a procedure which works only in special circumstances, he must be sure to include enough examples in which the procedure does not work to see that students do not overgeneralize.

4. Finally, the discovery sequence must include enough problems to provide practice and consolidation of the generalization.

Discovery teaching requires extremely careful planning by the teacher. In addition, it requires teacher patience to let students make their own connections and comparisons at their own pace. Gertrude Hendrix points out that most teachers who have been really successful with inductive methods have acquired the ability of making the learner *aware* of generalizations before calling for statements of those generalizations. (4:291) When the teacher pushes too quickly for conclusions, two things can happen. If the teacher pushes for generalizations before students have noticed any basic similarity among examples, the discussion becomes a guessing contest. Or, if pushed quickly, the generalization may be incomplete, incorrectly stated, or even false. Instead of driving for verbalization, the teacher should look for "nonverbal awareness." This awareness is signaled by the *action* of the learner when he is suddenly able to solve problems quickly and on his own or when he can give examples of the generalization without necessarily stating what that generalization is. Nonverbal awareness can have an electrifying effect on a classroom. Seeing that his neighbor has discovered a shortcut for doing problems is much more motivation for a student than instruction from the teacher to "look for shortcuts." This motivation and learning from peers produces a kind of bootstrap effect for active learning. Discovery learning is much more effective in a group situation where discussion can occur than in a student-teacher tutorial situation. For active discovery learning, optimal class size is *not* one!

Mathematics Laboratories and Activity Learning

Discovery learning is often related to inductive teaching techniques, but discovery learning also occurs with what could be termed situational teaching. This teaching strategy focuses on a particular situation or incident. Students are asked to analyze the situation mathematically and, in the course of this analysis, discover new mathematical generalizations and information. What sets situational teaching apart from a mere study of word problems is the use of physical materials. The student is given concrete materials which are so constructed that when they are combined and manipulated properly, mathematical relationships become apparent. Such teaching situations which rely on physical materials are often referred to as *mathematics laboratories*. (The term *laboratory* has been applied to a variety of educational situations recently, including individual learning "laboratories" which require no more than paper and pencil. However, for the purposes of this discussion, we will define *laboratory* as a situation requiring manipulation of a number of concrete materials.

What kind of physical apparatus lends itself to mathematical discovery? In many cases the best materials are also the simplest. For example, the graphic explanation of multiplication of binomials given in module 1-4 can be easily adapted for use in a mathematics laboratory. Cut three basic types of rectangular regions from light cardboard. The first region is a square one unit on a side. The second region is a larger square whose side dimensions are simply specified as "x." (Because students should not know the length of the side of the larger square in terms of the unit length, an arbitrary length should be chosen.) The third region is a rectangle whose width is one unit, and whose length is exactly "x," the length of the side of the larger square. Students are given an assortment of these three basic cardboard pieces, and asked to label the area of each piece. The areas will be 1, x^2, and $1x$ respectively. Students are then asked to make a rectangle whose dimensions are $(x+2)$ by $(x+3)$ by fitting the pieces of cardboard together, jigsaw fashion. Then they are asked to record the total area of the new rectangle by adding together the areas of the component pieces. They are instructed to repeat the procedure for several other specified rectangles, recording the total area in each case. After an inductive sequence of such manipulations, most children will quickly discover the pattern for multiplying binomials.

There are many advantages of the laboratory in this teaching strategy. First, it provides children with something to *do*. It makes learning an activity, and children are by their basic nature active. Second, most laboratories provide a way of minimizing the importance of computation during the initial stages of the discovery sequence by substituting materials for paper and pencil. As a result, a common stumbling block in mathematics is often delayed until the student has a good start on the problem. Finally, the mathematics laboratory makes clear the use of mathematics in the world of applications. By using familiar and "real" materials to generate mathematical abstractions, mathematics becomes "real" to the student.

Of course, incidental discovery learning in mathematics laboratories has its pitfalls just like ordinary discovery learning. All the important factors which determine the success or failure of an inductive sequence also operate in the laboratory. Most laboratories, like the one described in the example, depend upon carrying out manipulations (like fitting the cardboard pieces together) in a series of progressive situations which direct attention to changing aspects within the situation (like the varied dimensions of our rectangles). This series of situations is really an inductive sequence like the problem sheet discussed earlier, with manipulations substituted for paper-and-pencil calculations. The same sequencing conditions are again important, but they may take on slightly different forms in the laboratory.

1. The timing of laboratory experiences is crucial. The teacher must not only be sure that students have the necessary prerequisite mathematical skills in case computations are necessary in the course of the experiment, but he must also see that students have the necessary motor skills and familiarity with the apparatus to be able to correctly manipulate it. Experiences have shown, for example, that a laboratory exercise in compass-and-straight-edge geometric constructions is doomed to failure with first graders. What makes it a failure is *not* that they do not have the prerequisite mathematical understandings to do the geometrical constructions. It fails because most first graders have not yet developed motor skills which are finely controlled enough to allow them to accurately use a pair of compasses! This example is an extreme one, but it is one the teacher should keep in mind when considering a piece of apparatus which is complex and requires very fine adjustments, even though his students may be at the high school level.

2. The timing of laboratory experiences should emphasize discovery, not confirmation. With pencil-and-paper discovery sequences that the teacher should not state what is to be discovered before proceeding with the induction pattern seems obvious. Perhaps because the laboratory shows the applications of mathematics, this same restriction is not nearly as apparent there. It seems quite natural, in our binomial multiplication example, that confusion would be lessened if the teacher were to use the blackboard to explain the pattern between the factors and the product before ever handing out the cardboard squares. Such attempts to "speed up" mathematics laboratories by explaining the principles beforehand are the most common mistakes that teachers make in using the laboratory approach.

Why are such attempts to lessen confusion really mistakes? We can find the answer by considering the basic nature of mathematics itself. Mathematics is not a laboratory subject. The power and advantages of mathematics lie not in concrete manipulations, but in abstract symbolic manipulations, which is always the final goal that we seek. The final step in any laboratory experience should be the development of the power to carry out the calculation or to remember the structure *without reference to the physical objects*. But if the teacher *begins* by showing the mathematical principle without the use of the physical objects, what then is the reason for going to the laboratory? If we use the laboratory only to confirm an abstraction, we are forcing the student to move *backward* in his development since the mathematical abstraction is what the *final* goal should be.

Mathematics laboratories must be timed so they *introduce* mathematical situations and principles. This principle is a vital (and commonly overlooked) one for the successful use of mathematics laboratories.

3. Like discovery sequences, the best mathematics laboratories allow the student to reach an end result in more than one way. The cardboard squares and rectangles in the example can be fit together in many, many ways to form any of the required larger rectangles. When the student realizes that there are many approaches, he becomes willing to focus upon only the final area (product) and the original dimensions (factors) of the composite rectangle. If the result appeared to depend upon the pieces being fit together in only one way, there would not seem to be much reason to even seek a generalization. Thus, freedom of approach in a mathematics laboratory is important for broad mathematical generalizations.

Some teachers are reluctant to grant this freedom on the grounds that students may get the wrong answers and make a wrong generalization. However, to get a wrong answer in a laboratory is impossible. The difficulties that do arise come, not because mother nature has capriciously withheld the correct answer, but because the student has asked the wrong question. The following example may make this distinction clear.

Suppose we ask a student to form a rectangle that is $x+1$ by $x+2$. Instead of fitting the pieces together tightly, he fits them together loosely, using only unit squares and a larger x square. His final rectangle looks like Figure 16. When he writes down the product for his area, he has $(x+1)(x+2)=x^2+11$. Certainly that seems like a wrong answer. But it is actually a correct answer! The answer is correct *for the question which the student asked*. That question was, "What is the product of $(x+1)$ and $(x+2)$ equivalent to *if*

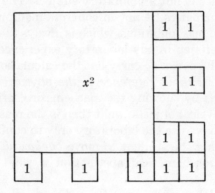

Figure 16. Discovery Sequence — Area of a Rectangle.

$x = 3$? When the student assumed that unit squares would fit along the side of the x square, he changed the question he was asking (even though he was not aware of it). The answer he arrived at is correct for his question. For if $x = 3$, then $(x + 1)(x + 2) = (4)(5) = 20$. And if $x = 3$, $x^2 + 11 = 9 + 11 = 20$. When "unusual" answers appear in a mathematics laboratory, the teacher should look for unusual aspects in the student's procedure which have, in effect, caused him to ask the wrong question.

4. A good discovery sequence must include enough cases for complete generalization. This requirement is also crucial for mathematics laboratories. Seeing a single example may convince the student that mathematics can be applied to that one instance, but it will almost always cause him to focus on the application rather than upon the mathematics. To emphasize the mathematics, he should see the principle in different physical embodiments. This perception is not always convenient or possible. In such cases, the teacher must see that enough varied cases using the same materials are presented to focus upon the mathematics rather than the materials.

Because of this requirement, a teacher demonstration is never the same as a mathematics laboratory. A demonstration seldom presents more than one or two situations at the most. Demonstrations also remove the opportunity for active learning. Displaying a fancy plastic model at the front of the classroom seldom encourages activity on the part of the learner!

5. A mathematics laboratory must generate *useful* mathematics. That is, the mathematical principles to be discovered with a

piece of apparatus should be important principles that are useful not only for mathematical applications, but for the development of additional mathematics itself. The mathematics laboratory should be able to fit into a larger overall teaching sequence. It should draw upon past mathematics which the student already knows. Our binomial multiplication example draws upon basic ideas about the measurement and calculation of area. Its introduction requires the student to review and use very important mathematical information already in his background. At the end of the experiment, the results can be checked by using the distributive principle, and the conclusions will be used again in the solution of quadratic equations. Good mathematics laboratories should be capable of being extended in just this way. Many gimmicks are being promoted as mathematics laboratories which are very interesting in themselves, but develop no mathematical ideas which reinforce past knowledge or suggest future mathematics. Good laboratory materials should do more than just provide for student activity. They should also be structured to promote the development of important and useful mathematical concepts.

A good example of physical materials which are structured to promote discovery of important mathematical concepts is the set of Cuisenaire rods named after their inventor Georges Cuisenaire. The set contains rods of ten different lengths (from one to ten centimeters in length). Each length has a specific color associated with it. The rods can be used to perform addition, subtraction, multiplication, and division operations. Since the rods bear no distinguishing markings except color, any rod can be designated as one. If the smallest rod is considered as one, operations with the natural numbers result. If a larger rod is taken as one, the set can be used to illustrate operations with fractions.

Addition and subtraction operations are performed by building "trains" of rods whose length is matched with a single rod or vice versa. Multiplication of two rods is done by making a cross with the two rods. Rods identical to the one on the bottom of the cross are placed until a "floor" has been formed under the top crossrod. The answer to the multiplication is then the sum of the "floor" rods. Since the lengths of Cuisenaire rods are multiples of their one centimeter width and one centimeter height, this definition of multiplication is equivalent to defining multiplication as repeated addition (that is, 3×2 means $3 + 3$; 3×5 means $3 + 3 + 3 + 3 + 3$; etc.) Similarly, division is taught as repeated subtraction.

With addition and multiplication defined as physical operations, students can also see if this definition includes the associative and commutative postulates. The rods can also be used to define odd, even, and prime numbers, to find the factors of a number, and even to explore permutations.

Today's mathematics classrooms are no longer limited to desks and blackboards. They are becoming filled with equipment which students can manipulate in a laboratory setting. Attribute block and logical blocks begin with problems of classification and grouping and develop such logical operations as conjunction and disjunction or union and intersection. The ancient abacus has been modified to an open-ended form where students place the beads on the rods. By varying the length of the rods, different number base counting systems can be illustrated. Informal geometry may be explored by stretching rubber bands over a lattice of nail heads to form various polygons on equipment known as geoboards. Construction kits of vinyl sticks and connectors extend this geometric exploration into three dimensions. But mathematics laboratories are not limited to commercial equipment. Some of the best laboratory materials are homemade components like sugar cubes, soda straws, and ping pong balls. The challenge of mathematics laboratories is a challenge to find mathematics in the everyday world around you and to help students see mathematics in this way, too.

For Further Investigation and Discussion

1. Piaget's research shows that children have concrete reasoning ability before they have abstract reasoning ability. Some people feel that *anytime* a new mathematical idea is introduced it should first be presented with physical materials (as in a mathematics laboratory) before it is presented abstractly. Do the stages of Piaget agree with this idea? Would you support the idea as a good teaching practice? Why or why not?

2. We do not teach mathematics in the same order or in the same detail in which it developed historically. Should we restrict ourselves to only teaching mathematics in the order indicated by Piaget's developmental studies? Would there ever be a reason for teaching an idea before an appropriate conservation stage had developed? (Is that even possible? Does your own school experience help answer these questions?)

3. Read Bruner, Jerome. "On Learning Mathematics." *The Mathematics Teacher* 53 (1960): 610-619. Compare his optimism about discovery learning with the views of David P. Ausubel in "Some Psychological and Educational Limitations of Learning by Discovery." *The Arithmetic Teacher* 11 (1964): 290-302. In what ways are these psychologists talking past each other? In what ways are their views complementary?

4. For techniques that are useful in discovery teaching read Hendrix, Gertrude. "Learning by Discovery." *The Mathematics Teacher* 54 (1961): 290-299. Another approach to discovery teaching techniques is found in Davis, Robert B. "Discovery in the Teaching of Mathematics." Chapter 8. *Learning by Discovery: A Critical Appraisal.* Edited by Lee S. Shulman and Evan R. Keislar. Chicago: Rand McNally & Co., 1966. Compare these two articles. Which of Davis' techniques might Hendrix *not* want to call discovery?

5. Read the article by Steffe, Leslie P. "Thinking About Measurement." *The Arithmetic Teacher* 18 (1971): 332-338. Outline a series of lessons for teaching young children to measure a book with a ruler using the suggestions made in this article.

For Lesson Planning

1. Obtain a recent issue of *The Arithmetic Teacher*. Develop a discovery worksheet from the examples contained in the "IDEAS" section. Evaluate your worksheet against the criteria listed in module 2-3.

2. Read the article, "An Annotated Bibliography of Suggested Manipulative Devices," by Patricia S. Davidson. *The Arithmetic Teacher* 15 (1968): 509-524. Make a list of the items which could be built from homemade sources like buttons, styrofoam cups, bottle caps, etc. For an example, see Higgins, Jon L. "Sugar-Cube Mathematics." *The Arithmetic Teacher* 16 (1969): 427-431.

3. Construct a manipulative device based upon any appropriate article in *The Arithmetic Teacher* or *The Mathematics Teacher*. Write out a lesson plan or description of how you would use it in the classroom.

4. Obtain a small set of Cuisenaire rods. Use the rods to demonstrate the Pythagorean theorem for the case of a 3-4-5 right triangle. Represent the squares of the sides by forming appropriate square surface areas with the rods. Fit the rods which form the squares of the legs on top of the rods which form the square of the hypothenuse. How many ways can this fitting or overlapping be done? Outline a sequence of lessons on the Pythagorean theorem which includes this activity.

For Microteaching

One of the most crucial aspects of discovery teaching and of the successful use of mathematics laboratories is the ability to give directions. Giving good directions is something like walking a tightrope. One must maintain a balance between too little information and too much information. Too much information may give away the point of the exercise and remove any reason for doing it. Too much information may simply provide more details than can be remembered and cause the student to only partially complete the task. The relative importance of many details may not be distinguished, which can add to confusion. Finally, the very act of giving too much information takes considerable time, and students often grow inattentive during the lag between motivation and action. When considerable detail seems absolutely necessary, it is wise to consider breaking the exercises into a series of smaller tasks so that directions can be given separately for each of them.

Giving too little information is, unfortunately, similar to giving too much information. If too few details are given, the student may again fail to see necessary intervening steps. If the goal is not made clear, then the importance of the details that are given may be hidden, and confusion about the entire exercise may result. Too little information may completely fail to motivate the exercise.

The art of giving good directions requires careful practice and analysis. Therefore, the microteaching technique is very helpful. Plan a microteaching lesson for your pupils which will require directions from you. You may want to construct a worksheet or a short laboratory activity like the examples given in module 2-3. Or you may find suggestions for an appropriate exercise in the section "For Lesson Planning" in this module. Outline carefully the major points you need to make in the directions you will give. Pay particular attention to the following three areas:

1. What is the purpose of doing this exercise? Does it solve a problem or surmount a difficulty in mathematics? Does it involve a useful mathematical procedure which should be practiced? Is it related to something similar that students have done before? Emphasizing any of these points will help students see the objective toward which your directions are pointing. You may find it helpful to encourage questions or allow discussion during this phase of your direction giving.

2. What are the important steps in the procedure? Can they be described without going through them, or would working out a sample be better? Do you need to emphasize certain steps as trickier or more important than others?

3. How can students tell when they are approaching the goal? Should they be able to do something very quickly? Should they be able to avoid a calculation by a short counting process? Try to give students some way to evaluate their own work without giving away the final answer or conclusion.

When you have planned your directions carefully, present them to a small group of five to eight students. Your presentation should be videotaped or audiotaped so that you can analyze it later. Unless discussion develops, your instructions should not take longer than two or three minutes.

When you have completed giving your directions, the recorders may be shut off. Walk around the room and look over students' shoulders. Are they doing what you expected them to do? Make notes about the unexpected things they may be doing or any hesitancy or confusion that develops. If time permits, allow students to finish the exercise. If your schedule is tight, you may observe for three or four minutes and then indicate that they can finish in another room or take the materials home.

Review any notes you took and try to categorize types of difficulties that you observed. Then view the videotape or listen carefully to the audiotape. Can you find things in your directions that might have led to misunderstandings? If not, ask someone else to view the videotape with you and give you his impression.

Write out a corrected version of your directions. If possible, evaluate the changes you made by repeating the microteach with a new group of students. And be patient! Giving good directions is a delicate art which requires not only practice but also careful observation and reflection.

For Related Research

Overviews of Piagetian Research

There is now considerable literature from Piaget and about Piaget available in English. Since some of it is difficult reading without the proper background, trying a specific program of reading is often wise. The following references suggest one such program.

Piaget's article, "Development and Learning," in the *Journal of Research in Science Teaching* 2 (1964): 176-186 is a good place to begin. A companion to this piece is the article by Eleanor Duckworth, "Piaget Rediscovered." *The Arithmetic Teacher* 11 (1964): 496-499.

A good introduction to Piaget's actual reports of many of his clinical studies is found in Bearley, Molly and Elizabeth Hitchfield. *A Teacher's Guide to Reading Piaget*. London: Routledge and Kegan Paul, 1966. This book reprints many of Piaget's reported interviews with children and organizes them in a sequence that is easier to follow than the original texts. You may want to contrast some of Piaget's original work with findings in similar areas of other psychologists who have recently tried to replicate many of Piaget's experiments. A good review and summary in this area is Harrison, D. B. "Piagetian Studies and Mathematics Learning." *Studies in Mathematics, Vol. XIX: Reviews of Recent Research in Mathematics Education*. Stanford, Calif.: School Mathematics Study Group, 1969. Harrison also gives a good overview and summary of Piaget's theories in the first section of his article. Another good overview is Beard, Ruth. *An Outline of Piaget's Developmental Psychology for Students and Teachers*. New York: Basic Books, 1969.

For applications to mathematics teaching, see the article by Irving Adler. "Mental Growth and the Art of Teaching." *The Arithmetic Teacher* 13 (1966): 576-584. Piaget's own interpretation of his work for school practice is found in Piaget, Jean. *Science of Education and the Psychology of the Child*. New York: The Viking Press, 1970.

Piaget explains the foundations of his work in the little book, *Genetic Epistemology*. New York: Columbia University Press, 1970. Of particular interest in this book are the parallels he draws between the structure of his developmental theory and the mathematical structure of Bourbaki. The original reports of Piaget that are most central to mathematics education are found in three books. They are:

Piaget, Jean. *The Child's Conception of Number.* New York: W. W. Norton & Co., 1965.

_____ and Barbel Inhelder. *The Child's Conception of Space.* New York: W. W. Norton & Co., 1967.

Inhelder, Barbel and Jean Piaget. *The Early Growth of Logic in the Child: Classification and Seriation.* New York: W. W. Norton & Co., 1969.

Specific Piagetian Research

Since the middle 1950's, Piagetian research has become an increasingly important part of mathematics education research. The scope of this section does not permit a complete review of this large body of research. Instead, we will suggest articles for reading which provide good introductions to different areas of Piagetian research. For an excellent overview of research results in this broad field read Lovell, Kenneth E. "Intellectual Growth and Understanding Mathematics." *Journal of Research for Mathematics Education* 3 (1972): 164-182.

Most of Piaget's original work was done with children in Geneva. An immediate question can be raised about the generality of his work, and this question makes replication studies very important. One such replication is reported in the article by David Elkind. "The Development of Quantitative Thinking: A Systematic Replication of Piaget's Studies." *Journal of Genetic Psychology* 98 (1961): 37-46. Additional replication in the area of seriation (serial ordering) is reported by Elkind. "Discrimination, Seriation, and Numeration of Size and Dimensional Differences in Young Children: Piagetian Replication Study VI." *Journal of Genetic Psychology* 104 (1964): 275-296. Additional replications in the area of geometry are reported in Laurendeau, Monique and Adrien Pinard. *The Development of the Concept of Space in the Child.* New York: International Universities Press, 1970. A replication study with children from a very different culture is Elizabeth Etuk's dissertation. *The Development of Number Concepts: An Examination of Piaget's Theory with Yoruba-Speaking Nigerian Children.* (Available from Dissertation Abstracts 28: 1295A, no. 4, 1967.)

Attempts are being made to determine factors that lead to achievement of conservation by children, in the hope that procedures for teaching or otherwise accelerating conservation can be inferred. One such experiment in the area of numerousness is reported in Wallach, Lise and R. L. Sprott. "Inducing Number Conservation in Children." *Child Development* 35 (1964): 1057-1072. A different approach to the same topic is reported by Wallach. "Number Conservation: The Roles of Reversibility, Addition, Subtraction, and Misleading Perceptual Cues." *Child Development* 38 (1967): 425-442. Another experiment in the area of seriation and ordering is reported in Coxford, Arthur. "The Effect of Instruction on the Stage Placement of Children in Piaget's Seriation Experiments." *The Arithmetic Teacher* 11 (1964): 4-9.

The relationship of children's ability in conservation tasks to their performance in school mathematics is another area of research concern. A study of conservation of numerousness as it relates to addition is reported in Van Engen, Henry and Leslie P. Steffe. *First Grade Children's Concept of Addition and Natural Numbers.* Madison: Wisconsin Research and Development Center for Cognitive Learning, Technical Report No. 5, 1966. An extension of this work is reported by Steffe in the article, "Relationships of Conservation of Numerousness to Problem-Solving Abilities of First Grade Children." *The Arithmetic Teacher* 15 (1968): 47-52. An English research study investigating the relationship between the child's development and his school achievement is reported in Hood, H. B. "An Experimental Study of Piaget's Theory of the Development of Number in Children." *British Journal of Psychology* 53 (1962): 273-286.

Research in training children to measure is reported in Beilen, Harry and I. C. Franklin. "Logical Operations in Area and Length Measurement: Age and Training Effects." *Child Development* 33 (1962): 607-618. The relationship between such research and the school mathematics curriculum is discussed in Sawada, Daiyo, and L. Doyal Nelson. "Conservation of Length and the Teaching of Linear Measurement: A Methodological Critique." *The Arithmetic Teacher* 14 (1967): 345-348. Another analysis of this area is found in Huntington, Jefferson R. "Linear Measurement in the Primary Grades: A Comparison of Piaget's Description of the Child's Spontaneous Conceptual Development and the SMSG Sequence of Instruction." *Journal for Research in Mathematics Education* 1 (1970): 219-232.

Finally, Piagetian research has suggested developmental research in curriculum building. These efforts have not been concerned with problems of acceleration so much as they have been with the development and evaluation of school mathematics units that are consistent with Piaget's theories. An example of such work is found in Shah, S. A. "Selected Geometric Concepts Taught to Children Ages Seven to Eleven." *The Arithmetic Teacher* 16 (1969): 119-128.

References to Unit 2

1. Berlyne, D. E. "Recent Developments in Piaget's Work." *British Journal of Educational Psychology* 27 (1957): 1-12.
2. Flavell, John. *The Developmental Psychology of Jean Piaget.* Princeton, N. J.: D. Van Nostrand Co., 1963.
3. Harrison, D. B. "Piagetian Studies and Mathematics Learning." *Studies in Mathematics Volume XIX: Review of Recent Research in Mathematics Education.* Stanford, Calif.: School Mathematics Study Group, 1969.
4. Hendrix, Gertrude. "Learning by Discovery." *The Mathematics Teacher* 54 (1961): 290-299.
5. Jowlett, Benjamin, trans. *The Dialogs of Plato*; vol. I. New York: Random House, 1937.
6. Karplus, Elizabeth F. and Robert Karplus. "Intellectual Development Beyond Elementary School, 1: Deductive Logic." *School Science and Mathematics* 70 (1970): 398-407.
7. Piaget, Jean. "Development and Learning." *Journal of Research in Science Teaching* 2 (1964): 176-186.
8. _____. "How Children Form Mathematical Concepts." *Scientific American* 189 (1953): 74-75ff.

Unit 3

Stimulus-Response Learning in Mathematics

Introduction

We dare not limit ourselves to a study of different abilities to learn mathematics. Teachers must be even more interested in how people learn mathematics for how the teacher teaches and how the student learns should be intimately related. Since you have done much more than an average amount of mathematics learning, you may already know how mathematics is learned. You should weigh the theories discussed in these units in terms of your own learning experiences. At the same time, however, you should expect to find more than what just common sense would reveal. Everyone goes through most of the stages of Piaget in the course of becoming an adult, yet hardly any adult remembers these different ways of responding or even anticipates differentiated responses from children. Similarly, despite the fact that you have done a great deal of mathematics learning, you have probably not been cognizant of the processes you were employing in such learning. Therefore, looking at the way people respond to different learning tasks before examining psychological theories of learning as they apply to mathematics is essential. This unit begins with such an investigation and presents an accompanying theory in a strong and positive manner. You should carry out the investigation and read the following modules with a healthy sense of skepticism. In subsequent units we will present other investigations and other theories which develop quite different analyses of learning. Mathematics learning is very complex. No single learning theory can satisfactorily handle this complexity, yet each learning theory contributes new understandings of how children learn mathematics.

Stimulus-Response Pairs and Meaningful Learning

In a number of learning situations, the student must develop the ability to give a particular verbal response to a particular verbal cue. Learning a list of pairs is called paired-associate learning. In this investigation we shall examine the role of meaningfulness in such learning. To completely understand the concept of meaningfulness requires both philosophical as well as psychological considerations. For the purposes of this investigation let us agree that word "A" is more meaningful than word "B" if a group of subjects can, on the average, write down more associations to word "A" than they can to word "B" in a given time period. Beyond this condition, we shall not consider relative degrees of meaningfulness but shall confine ourselves to extremes by comparing two-syllable mathematics vocabulary words with two-syllable nonsense words.

For the investigation we will present the subject with eight pairs of words. After a study trial, we will present the first word of the pair (stimulus) and ask the subject to try to give the second word of the pair (response). We will repeat this procedure through the list for a total of fifteen attempts for each pair. The list consists of four types of pairings:

1. High stimulus, high response meaningfulness
2. High stimulus, low response meaningfulness
3. Low stimulus, high response meaningfulness
4. Low stimulus, low response meaningfulness

We will attempt to answer two questions:

    1. Will high *stimulus* items be learned in fewer trials than low stimulus items?
    2. Will high *response* items be learned in fewer trials than low response items?

You should realize that these two questions are conceptually independent. That is, if one is answered affirmatively, it is still entirely possible for the other to be answered negatively.

Here is our list of word pairs. Prepare for the experiment by printing each pair on one side of an index card. Write the number of the card in the upper right-hand corner. On the opposite side of the card write only the first word of the pair. (If you use a felt pen, be sure that markings do not show through the opposite sides of the cards.

1st card:	INVERSE	— EQUAL
2nd card:	ANGLE	— RATIO
3rd card:	MEARDON	— ZUHAP
4th card:	BYSSUS	— POLEF
5th card:	DIVIDE	— NARON
6th card:	SEGMENT	— GOJEY
7th card:	VOLVAP	— SUBTRACT
8th card:	SAGROLE	— FRACTION

To be sure that the important association in the learning task is between each word pair and not with the order of the list, the cards must be presented to the subject in a different order each trial. This order has been specified on the following scoring sheet. You must rearrange the cards between each trial. Give the subject four seconds to respond to each card. Mark the response correct (+), incorrect (−), or not given (0). Remember that for the study trial, the side of the card showing both words is shown to the subject (and he reads them aloud). For all other trials, the side containing the stimulus word only is presented. (Since you will be able to see the back side of the card, you can immediately judge if the subject gave the correct response.) If the response is incorrect, turn the card over to show the correct pairing. Allow thirty seconds between trials. Do not omit any trials, even though the responses on the first trial may all be correct.

Study Trial:	card no.	5	4	2	7	6	8	3	1
	response	x	x	x	x	x	x	x	x
Trial 1:	card no.	4	3	2	7	1	6	5	8
	response								
Trial 2:	card no.	7	3	5	1	4	2	6	8
	response								
Trial 3:	card no.	1	4	5	7	6	8	3	2
	response								
Trial 4:	card no.	5	7	1	4	2	8	3	6
	response								
Trial 5:	card no.	8	2	4	7	5	1	6	3
	response								
Trial 6:	card no.	2	4	7	5	8	1	3	6
	response								
Trial 7:	card no.	3	8	2	5	4	1	7	6
	response								
Trial 8:	card no.	5	1	3	7	6	4	8	2
	response								
Trial 9:	card no.	1	7	4	6	8	3	2	5
	response								
Trial 10:	card no.	3	6	8	1	4	2	7	5
	response								
Trial 11:	card no.	8	6	7	2	4	5	3	1
	response								
Trial 12:	card no.	6	4	7	1	3	5	2	8
	response								
Trial 13:	card no.	4	1	8	6	7	2	3	5
	response								
Trial 14:	card no.	1	6	7	4	5	8	2	3
	response								
Trial 15:	card no.	7	1	8	6	4	2	3	5
	response								

Table 7. Subject Response in Stimulus-Response Test.

Analyzing the Data

Complete the following summary sheet from the scores you recorded during the experiment.

Trial Number	Number of Correct Responses (+)	Number of Omissions (0)	Number of Errors (-)
1.			
2.			
3.			
4.			
5.			
6.			
7.			
8.			
9.			
10.			
11.			
12.			
13.			
14.			
15.			

Table 8. Summary Sheet of Student Response.

You can show the learning progress visually by plotting the number of correct responses against the trial number. It is common practice to join the plotted points with a curve. Such a curve is commonly called a *learning curve*. Compare the learning curve of your subject with the curves for other subjects measured by other members of the class and discuss discrepancies. Are these discrepancies due primarily to differences in ability or to differences in the way the experiment was presented? Save your learning curves for use later in the chapter.

To answer our original questions we need a different kind of summary according to the meaningfulness of the items. Record the total number of correct responses, omissions, and errors for each item across all trials in the following table.

Item	Number of Correct Responses (+)	Number of Omissions (0)	Number of Errors (-)
1. Inverse: Equal			
2. Angle: Ratio			
3. Meardon: Zuhap			
4. Byssus: Polef			
5. Divide: Naron			
6. Segment: Gojey			
7. Volvap: Subtract			
8. Sagrole: Fraction			
High: High (1+2)			
Low: Low (3+4)			
High: Low (5+6)			
Low: High (7+8)			

Table 9. Summary of Responses by Type of Stimulus.

Were high *stimulus* items learned in fewer trials than low stimulus items? Were high response items learned in fewer trials than low response items? How can you tell whether your differences are due to differences in learning or to chance occurrence?

Differences Between Measures

From your data find the total number of correct responses your subject made in 15 trials to the high stimulus items (items 1, 2, 5, and 6). Similarly, find the total number of correct responses made to low stimulus items (items 3, 4, 7 and 8). Share your totals with other class members so that you can consider this data for a minimum of ten subjects. (If this is not possible, you will need to recruit more subjects.) You should be able to complete a table like this:

Subject Number	Correct Responses, High Stimulus (H)	Correct Responses, Low Stimulus (L)	Difference D = H - L
1			
2			
3			

Table 10. Differences in High-Stimulus and Low-Stimulus Responses.

Compute a difference score for each subject and find the mean of these difference scores, $\overline{D}$. Except in unusual cases, this mean difference score will not be zero. Is this because there is a difference between learning high stimulus and low stimulus items, or because it is difficult to find a case where these differences exactly eliminate each other?

Imagine that we had administered this experiment to an infinite number of people and that the ten (or more) cases you are considering represents a sample of that population. That the mean of the sample considered would not be the same as the mean of the entire population, u, is possible. In fact, if we randomly selected another sample, its mean would probably not be quite the same as the mean of the first sample or as the population mean. Suppose that we did select a very large number of samples of our population, computing the mean for each. We could look at the distribution of these sample means. We should find that most of the sample means have a value close to that of the population mean. Although there will be some samples whose means will be quite different from the population mean, the number of such sample means should decline drastically as the difference increases. As this difference, x, increases, the probability of selecting a sample whose mean is greater than $(u+x)$ or less than $(u-x)$ becomes smaller and smaller. For normally distributed population scores, this probability can be theoretically computed. It is possible to form a table of difference values, x, vs. the probability, p, of finding a sample mean *outside* the range $(u-x)$ to $(u+x)$. If we found a value of $p=0.05$ for an x value of 1.96, it would mean that the probability of finding a sample mean greater than $u+1.96$ or less than $u-1.96$ was less than 5 chances in 100.

There is one "catch" in this procedure which may have already occurred to you. The size of x also depends upon the size of the scores expected from our experiment. To solve this problem, we consider the distribution of the ratio

$$t = \frac{\overline{X}_s - u}{\sigma_{\overline{X}}}$$

where $\overline{X}_s$ is the sample mean, u is the population mean, and $\sigma_{\overline{X}}$ is the standard deviation of the distribution of the sample means (commonly called the *sample error*). For any sample, our ratio t expresses how far away its mean is from the population mean in standard error units. The size of the standard error unit depends upon the size of the standard deviation; it can be shown to be $\sigma/\sqrt{N}$ which in turn depends upon the size of the raw scores. Since the difference in the means also depends upon the size of the raw scores, the effect of the ratio t is to eliminate this factor. We can compare the distributions of t for different experiments.

In most cases, the standard deviation of the total population is not known, and the sample error is estimated by using the standard deviation of the sample instead. In this case the value of t for a given sample becomes

$$t = \frac{\overline{X}_s - u}{(\sigma/\sqrt{N})} = \frac{\overline{X}_s - u}{\sqrt{\dfrac{\sum(\overline{X}_s - X_i)^2}{N-1}}\Big/\sqrt{N}}$$

or

$$t = \frac{\overline{X}_s - u}{\sqrt{\dfrac{\sum(\overline{X}_s - X_i)^2}{N(N-1)}}}$$

where $\overline{X}_s$ is the mean of the sample, X_i are the sample scores, N is the size of the sample, and u is the population mean. The nature of this t also depends upon the size of the sample, N. Therefore, to locate the probability of a specific t value, we must also know the size of the sample, N. Distributions have been calculated for many values of N. They are commonly tabulated in terms of degrees of freedom, which in our case is N-1.

We will use a computation of t to see if there are significant differences between the learning of high stimulus and low stimulus items in the following way:

1. We will work with the difference scores you have already obtained.

2. *Assume* that, on the average, there is *no* difference. Then on a large population the mean difference score should be zero ($u = o$ by assumption).

3. Consider your subjects to be a sample of this larger, no-difference population. Compute their t-ratio. Since we have assumed $u = o$, their t-ratio is given by

$$t = \frac{\overline{X}_s - 0}{\sqrt{\dfrac{\sum(\overline{X}_s - X_i)^2}{N(N-1)}}} = \frac{\overline{D}}{\sqrt{\dfrac{\sum(\overline{D} - D_i)^2}{N(N-1)}}}$$

(We have substituted D's for X's to emphasize that the scores we are working with are *differences.*)

4. We now want to find whether there is a high probability or a low probability that this value of t could have occurred by chance. Use the table, "Critical Values of T," in the appendix. In the first column (df, Degrees of Freedom) locate the value which corresponds to $(N\text{-}1)$. Read this line across until you find the column containing the largest critical value which your t-value still exceeds. The number at the head of this column (Level of Significance) is the probability that your t-ratio could have occurred by chance.

5. Suppose the probability value you found in step 4 is high — 0.20, or 20 chances in 100. Then the chances of selecting a sample of subjects from the no-difference population whose t-value matches your sample is relatively high. That is, the differences you found are not significantly different from differences that would be found by sampling a no-difference population.

Now suppose that the probability value you found in step 4 was low — 2 chances in 100. Then the chances of matching your subjects by a random sample from a no-difference population are slim indeed. Your subjects were *not* learning as if no differences existed between high stimulus and low stimulus items.

Once you have gone through this procedure it becomes fairly easy. Use it again to test whether there was a significant difference in your subjects between:

1. high response versus low response.
2. high stimulus and response vs. low stimulus and response.

Do your results match your predictions? This experiment often goes counter to predictions. That is, many subjects often learn the nonsense syllables first. When they do, we cannot simply assume that our subjects have acted incorrectly. Perhaps we have been asking the "wrong question" by structuring the experiment incorrectly. Are the nonsense syllables really meaningless? Could they serve as attention getters? If so, could they be meaningful in another sense or dimension?

A Meaningful Teaching Experiment

You probably saw in investigation 3-1 that meaning could play a role in learning tasks although the exact nature of that role may be surprising. Of course, the learning task used for the experiment was not really typical of the vast majority of learning tasks encountered in a mathematics classroom. Our task was purposely simplified to emphasize a meaningful-nonsense dichotomy. This dichotomy may have had the effect of endowing the nonsense words with a special "meaning" of their own. In the classroom, such differences are never so exaggerated, but the task is much more practical. Can differences between meaningful and rote learning be seen in these situations?

The classical study in meaningful arithmetic learning was done in 1942-43 by William A. Brownell and Harold E. Moser (1). Approximately 1300 third graders were taught subtraction by either a mechanical (rote) or a meaningful (rational) process. At the time of the study, regrouping techniques in the subtraction process were commonly introduced in the third grade. Brownell arranged to have some classes taught on "equal additions process" and others taught a "minuend regrouping" process. Both processes may be taught by rote procedures as well as by meaningful explanations. The rote procedures are suggested by the examples below:

$$
\begin{array}{cc}
3\ 4\ ^{1}6 & 3\ ^{3}\!\!\!\!4\ ^{1}6 \\
2\ ^{4}\!\!\!5\ 9 & 2\ 3\ 9 \\
\hline
1\ 0\ 7 & 1\ 0\ 7
\end{array}
$$

equal additions   minuend regrouping

In the "equal additions" procedure, 10 is added to 346 to form 340+16. An equal amount (10) is also added to 239 to form 249. The "minuend regrouping" simply involves rewriting 346 as 330+16.

Classes were given tests of prerequisite mathematical abilities (addition and simple subtraction) as well as IQ tests before the teaching period began. At the end of the experimental period (three weeks) children were given a computation test in subtraction and were interviewed individually. These tests supplied measures of accuracy, rate of work, and transfer to new situations, while the interviews checked work habits and levels of understanding. Another computation test was given six weeks later as a measure of retention, and children were reinterviewed to check any changes of work habits.

The report of the experiment gives a thorough account of the teaching instructions given to teachers and of the testing and interviewing procedures followed, as well as the statistical computations employed. Brownell found that, in general, students taught by meaningful methods were superior on the computation test at the end of the treatment to students who had been taught by mechanical methods. These meaningfully-taught students remained superior in their ability to subtract when given a retention test several weeks after the experiment. He also found (although less clearly) that minuend regrouping resulted in higher student scores than the equal additions process.

It is interesting to note that textbook authors immediately took Brownell's study as an endorsement of the superiority of meaningful learning for all arithmetic processes at *all* grade levels despite the fact that the report carefully acknowledges the limitations and practical difficulties of the study. In a sense, this research was accepted as proof of a premise which, in fact, had already been accepted — a premise that held meaning to be so important in mathematics that it could not be ignored as a factor in the process of learning. Actually, the Brownell-Moser experiment did not focus upon the actions of the learner, but upon the procedures and explanations given by the classroom teacher. The experiment should be considered as one about meaningful teaching as much as about meaningful learning. The experiment suggests that for this particular task (subtraction) at this particular grade level (third grade), students' performances will be better when the students have access to not only an explanation of the process but also to a justification of the process according to basic properties of numbers.

This conclusion is a very important one that we must be aware of as we are teachers. However, it does not explain how students process information which is provided by the teacher. To understand learning we need to explore connections between the information or stimulus given by a teacher and the responses given by students. Look again at the results of investigation 3-1 to see if they suggest any such connections.

Interpreting the Investigation

How did your subjects learn how to make the correct responses? Which aspects of the investigation facilitated his learning? Which aspects made that learning more difficult? Answering these questions is a small, but important, step in knowing how to design a classroom teaching procedure.

Begin by looking at the end of investigation 3-1. How did you know that the subject had learned? How did you know whether to mark a +, −, or 0 in the score table? The obvious answer is that you judged whether learning had occurred by matching the response (action) of the subject against a predetermined criterion. In one sense, this matching is the essence of education. An educated man should respond, act, or behave differently than an uneducated man. And these responses should match established criteria; that is, he should act in constructive and desirable ways. But developing the response criteria in the case of an "educated man" is certainly much different than determining the response criteria in the case of the word-pair investigation.

This active view of learning and education is known as behaviorism. It is in part an outgrowth of the viewpoint established with so much success by American psychologists at the beginning of the twentieth century. The first psychological laboratory was established in Germany in 1879 by Wilhelm Wundt. Wundt and his colleagues were largely interested in understanding conscious experience in terms of basic components. They concentrated on relationships between man's sensations, thoughts, and feelings. Perhaps this interest in part reflected the romanticism of the era. But American psychologists were rooted in a developing tradition of pragmatism. Hence, their interests lay in practical usefulness, objective behavior, and the study of what people do.

What people do, according to such American psychologists as Watson, Thorndike and Skinner is *respond* to stimuli. Certainly, viewing investigation 3-1 in these terms is easy. The subject was presented with a series of stimuli and a set of appropriate responses during the study trial. What was learned during subsequent trials was to give the appropriate response to each stimulus. Perhaps you noticed one trial in which the subject gave the correct response to a stimulus word only to be followed on the next trial by the wrong response to the same stimulus word. Learning is not an all-or-nothing situation, but a task which develops over time.

The American psychologist Edward L. Thorndike (1874-1949) concluded that learning was the gradual "stamping-in" of stimulus-response connections. Furthermore, the stamping-in of stimulus-response connections depends not only on the fact that the stimulus and the response occurred together but also on the effects that follow the response. According to Thorndike, if a stimulus is followed by a response and then by a *satisfier*, the stimulus-response connection is strengthened. If, on the other hand, a stimulus is followed by a response and then by an *annoyer*, the stimulus-response connection is weakened. It is easy to visualize satisfiers and annoyers in terms of physical rewards and punishments given to animals in the laboratory. But what could be acting as a reward or punishment in investigation 3-1? The answer is provided by B. F. Skinner. The punishment (or negative reinforcement, to use Skinner's term) comes when the card is turned over and the subject sees that he has made the wrong response. The positive reinforcement is the pleasure of discovering that he has made the correct response.

The Science of Learning and the Art of Teaching[1]

B. F. Skinner

Some promising advances have recently been made in the field of learning. Special techniques have been designed to arrange what are called "contingencies of reinforcement" — the relations which prevail between behavior on the one hand and the consequences of that behavior on the other — with the result that a much more effective control of behavior has been achieved. It has long been argued that an organism learns mainly by producing changes in its environment, but it is only recently that these changes have been carefully manipulated. In traditional devices for the study of learning — in the serial maze, for example, or in the T-maze, the problem box, or the familiar discrimination apparatus — effects produced by the organism's behavior are left to many fluctuating circumstances. There is many a slip between the turn-to-the-right and the food-cup at the end of the alley. It is not surprising that techniques of this sort have yielded only very rough data from which the uniformities demanded by an experimental science can be extracted only by averaging many cases. In none of this work has the behavior of the individual organism been predicted in more than a statistical sense. The learning processes which are the presumed object of such research are reached only through a series of inferences. Current preoccupation with deductive systems reflects this state of the science.

Recent improvements in the conditions which control behavior in the field of learning are of two principal sorts. The Law of Effect has been taken seriously; we have made sure that effects *do* occur and that they occur under conditions which are optimal for producing the changes called learning. Once we have arranged the particular type of consequence called a reinforcement, our techniques permit us to shape up the behavior of an organism almost at will. It has become a routine exercise to demonstrate this in classes in elementary psychology by conditioning such an organism as a pigeon. Simply by presenting food to a hungry pigeon at the right time, it is possible to shape up three or four well-defined responses in a single demonstration period—such responses as turning around, pacing the floor in the pattern of a figure-8, standing still in a corner of the demonstration apparatus, stretching the neck, or stamping the foot. Extremely complex performances may be reached through successive stages in the shaping process, the contingencies of

Reprinted from B. F. Skinner, "The Science of Learning and the Art of Teaching," *Harvard Educational Review*, 24, Spring 1954, 86-97. Copyright © 1954 by the President and Fellows of Harvard College.

[1]Paper presented at a conference on Current Trends in Psychology and the Behavioral Sciences at the University of Pittsburgh, March 12, 1954.

reinforcement being changed progressively in the direction of the required behavior. The results are often quite dramatic. In such a demonstration one can *see* learning take place. A significant change in behavior is often obvious as the result of a single reinforcement.

A second important advance in technique permits us to maintain behavior in given states of strength for long periods of time. Reinforcements continue to be important, of course, long after an organism has learned *how* to do something, long after it has acquired behavior. They are necessary to maintain the behavior in strength. Of special interest is the effect of various schedules of intermittent reinforcement. Charles B. Ferster and the author are currently preparing an extensive report of a five-year research program, sponsored by the Office of Naval Research, in which most of the important types of schedules have been investigated and in which the effects of schedules in general have been reduced to a few principles. On the theoretical side we now have a fairly good idea of why a given schedule produces its appropriate performance. On the practical side we have learned how to maintain any given level of activity for daily periods limited only by the physical exhaustion of the organism and from day to day without substantial change throughout its life. Many of these effects would be traditionally assigned to the field of motivation, although the [principal operation is simply the arrangement of contingencies of reinforcement.][2]

These new methods of shaping behavior and of maintaining it in strength are a great improvement over the traditional practices of professional animal trainers, and it is not surprising that our laboratory results are already being applied to the production of performing animals for commercial purposes. In a more academic environment they have been used for demonstration purposes which extend far beyond an interest in learning as such. For example, it is not too difficult to arrange the complex contingencies which produce many types of social behavior. Competition is exemplified by two pigeons playing a modified game of ping-pong. The pigeons drive the ball back and forth across a small table by pecking at it. When the ball gets by one pigeon, the other is reinforced. The task of constructing such a "social relation" is probably completely out of reach of the traditional animal trainer. It requires a carefully designed program of gradually changing contingencies and the skillful use of schedules to maintain the behavior in strength. Each pigeon is separately prepared for its part in the total performance, and the "social relation" is then arbitrarily constructed. The sequence of events leading up to this stable state are excellent material for the study of the factors important in nonsynthetic social behavior. It is instructive to consider how a similar series of contingencies could arise in the case of the human organism through the evolution of cultural patterns.

Cooperation can also be set up, perhaps more easily than competition. We have trained two pigeons to coordinate their behavior in a cooperative endeavor with a precision which equals that of the most skillful human dancers. In a more serious vein these techniques have permitted

---

[2]The reader may wish to review Dr. Skinner's article, "Some Contributions of an Experimental Analysis of Behavior to Psychology as a Whole," *The American Psychologist*, 1953, 8, 69-78. Ed.

us to explore the complexities of the individual organism and to analyze some of the serial or coordinate behaviors involved in attention, problem solving, various types of self-control, and the subsidiary systems of responses within a single organism called "personalities." Some of these are exemplified in what we call multiple schedules of reinforcement. In general a given schedule has an effect upon the rate at which a response is emitted. Changes in the rate from moment to moment show a pattern typical of the schedule. The pattern may be as simple as a constant rate of responding at a given value, it may be gradually accelerating rate between certain extremes, it may be an abrupt change from not responding at all to a given stable high rate, and so on. It has been shown that the performance characteristic of a given schedule can be brought under the control of a particular stimulus and that different performances can be brought under the control of different stimuli in the same organism. At a recent meeting of the American Psychological Association, Dr. Ferster and the author demonstrated a pigeon whose behavior showed the pattern typical of "fixed-interval" reinforcement in the presence of one stimulus and, alternately, the pattern typical of the very different schedule called "fixed ratio" in the presence of a second stimulus. In the laboratory we have been able to obtain performances appropriate to *nine* different schedules in the presence of appropriate stimuli in random alternation. When Stimulus 1 is present, the pigeon executes the performance appropriate to Schedule 1. When Stimulus 2 is present, the pigeon executes the performance appropriate to Schedule 2. And so on. This result is important because it makes the extrapolation of our laboratory results to daily life much more plausible. We are all constantly shifting from schedule to schedule as our immediate environment changes, but the dynamics of the control exercised by reinforcement remain essentially unchanged.

It is also possible to construct very complex *sequences* of schedules. It is not easy to describe these in a few words, but two or three examples may be mentioned. In one experiment the pigeon generates a performance appropriate to Schedule A where the reinforcement is simply the production of the stimulus characteristic of Schedule B, to which the pigeon then responds appropriately. Under a third stimulus, the bird yields a performance appropriate to Schedule C where the reinforcement in this case is simply the production of the stimulus characteristic of Schedule D, to which the bird then responds appropriately. In a special case, first investigated by L. B. Wyckoff, Jr., the organism responds to one stimulus where the reinforcement consists of the *clarification* of the stimulus controlling another response. The first response becomes, so to speak, an objective form of "paying attention" to the second stimulus. In one important version of this experiment, as yet unpublished, we could say that the pigeon is telling us whether it is "paying attention" to the *shape* of a spot of light or to its *color*.

One of the most dramatic applications of these techniques has recently been made in the Harvard Psychological Laboratories by Floyd Ratliff and Donald S. Blough, who have skillfully used multiple and serial schedules of reinforcement to study complex perceptual processes in the infrahuman organism. They have achieved a sort of psycho-physics without verbal instruction. In a recent experiment by Blough, for example, a pigeon draws a detailed dark-adaptation curve showing the characteristic breaks of rod and cone vision. The curve is recorded continuously in a single experimental period and is quite comparable with the curves of human subjects. The pigeon behaves in a way which, in the human case, we would not hesitate to describe by saying that it adjusts a very faint patch of light until it can just be seen.

In all this work, the species of the organism has made surprisingly little difference. It is true that the organisms studied have all been vertebrates, but they still cover a wide range. Comparable results have been obtained with pigeons, rats, dogs, monkeys, human children, and most recently, by the author in collaboration with Ogden R. Lindsley, human psychotic subjects. In spite of great phylogenetic differences, all these organisms show amazingly similar properties of the learning process. It should be emphasized that this has been achieved by analyzing the effects of reinforcement and by designing techniques which manipulate reinforcement with considerable precision. Only in this way can the behavior of the individual organism be brought under such precise control. It is also important to note that through a gradual advance to complex interrelations among responses, the same degree of rigor is being extended to behavior which would usually be assigned to such fields as perception, thinking, and personality dynamics.

From this exciting prospect of an advancing science of learning, it is a great shock to turn to that branch of technology which is most directly concerned with the learning process — education. Let us consider, for example, the teaching of arithmetic in the lower grades. The school is concerned with imparting to the child a large number of responses of a special sort. The responses are all verbal. They consist of speaking and writing certain words, figures, and signs which, to put it roughly, refer to numbers and to arithmetic operations. The first task is to shape up these responses — to get the child to pronounce and to write responses correctly, but the principal task is to bring this behavior under many sorts of stimulus control. This is what happens when the child learns to count, to recite tables, to count while ticking off the items in an assemblage of objects, to respond to spoken or written numbers by saying "odd," "even," "prime," and so on. Over and above this elaborate repertoire of numerical behavior, most of which is often dismissed as the product of rote learning, the teaching of arithmetic looks forward to those complex serial arrangements of responses involved in original mathematical thinking. The child must acquire responses of transposing, clearing fractions, and so on, which modify the order pattern of the original material so that the response called a solution is eventually made possible.

Now, how is this extremely complicated verbal repertoire set up? In the first place, what reinforcements are used? Fifty years ago the answer would have been clear. At that time educational control was still frankly aversive. The child read numbers, copied numbers, memorized tables, and performed operations upon numbers to escape the threat of the birch rod or cane. Some positive reinforcements were perhaps eventually derived from the increased efficiency of the child in the field of arithmetic and in rare cases some automatic reinforcement may have resulted from the sheer manipulation of the medium — from the solution of problems or the discovery of the intricacies of the number system. But for the immediate purposes of education the child acted to avoid or escape punishment. It was part of the reform movement known as progressive education to make the positive consequences more immediately effective, but any one who visits the lower grades of the average school today will observe that a change has been made, not from aversive to positive control, but from one form of aversive stimulation to another. The child at his desk, filling in his workbook, is behaving primarily to escape from the threat of a series of minor aversive events — the teacher's displeasure, the criticism or ridicule of his classmates, an ignominious showing in a competition, low marks, a trip to the office "to be talked to" by the principal, or a word to the parent who may still resort to the birch rod. In this welter of aversive consequences, getting the right answer is in itself an insignificant event, any effect of which is lost amid the anxieties, the boredom, and the aggressions which are the inevitable by-products of aversive control.[3]

Secondly, we have to ask how the contingencies of reinforcement are arranged. When is a numerical operation reinforced as "right"? Eventually, of course, the pupil may be able to check his own answers and achieve some sort of automatic reinforcement, but in the early stages the reinforcement of being right is usually accorded by the teacher. The contingencies she provides are far from optimal. It can easily be demonstrated that, unless explicit mediating behavior has been set up, the lapse of only a few seconds between response and reinforcement destroys most of the effect. In a typical classroom, nevertheless, long periods of time customarily elapse. The teacher may walk up and down the aisle, for example, while the class is working on a sheet of problems, pausing here and there to say right or wrong. Many seconds or minutes intervene between the child's response and the teacher's reinforcement. In many cases — for example, when papers are taken home to be corrected — as much as 24 hours may intervene. It is surprising that this system has any effect whatsoever.

A third notable shortcoming is the lack of a skillful program which moves forward through a series of progressive approximations to the final complex behavior desired. A long series of contingencies is necessary to bring the organism into the possession of mathematical behavior most efficiently. But the teacher is seldom able to reinforce at each step in such a series because she cannot deal with the pupil's responses one at a time. It is usually necessary to reinforce the behavior in blocks of responses — as in correcting a work sheet or page from a workbook. The responses within such a block must not be interrelated. The answer to one problem must not depend upon the answer to another. The number of stages through which one may progressively approach a complex pattern of behavior is therefore small, and the task so much the more difficult. Even the most modern workbook in beginning arithmetic is far from exemplifying an efficient program for shaping up mathematical behavior.

Perhaps the most serious criticism of the current classroom is the relative infrequency of reinforcement. Since the pupil is usually dependent upon the teacher for being right, and since many pupils are usually dependent upon the same teacher, the total number of contingencies which may be arranged during, say, the first four years, is of the order

[3]Skinner, B. F. *Science and Human Behavior*. New York: Macmillan, 1953.

of only a few thousand. But a very rough estimate suggests that efficient mathematical behavior at this level requires something of the order of 25,000 contingencies. We may suppose that even in the brighter student a given contingency must be arranged several times to place the behavior well in hand. The responses to be set up are not simply the various items in tables of addition, subtraction, multiplication, and division; we have also to consider the alternative forms in which each item may be stated. To the learning of such material we should add hundreds of responses concerned with factoring, identifying primes, memorizing series, using short-cut techniques of calculation, constructing and using geometric representations or number forms, and so on. Over and above all this, the whole mathematical repertoire must be brought under the control of concrete problems of considerable variety. Perhaps 50,000 contingencies is a more conservative estimate. In this frame of reference the daily assignment in arithmetic seems pitifully meagre.

The result of all this is, of course, well known. Even our best schools are under criticism for their inefficiency in the teaching of drill subjects such as arithmetic. The condition in the average school is a matter of wide-spread national concern. Modern children simply do not learn arithmetic quickly or well. Nor is the result simply incompetence. The very subjects in which modern techniques are weakest are those in which failure is most conspicuous, and in the wake of an ever-growing incompetence come the anxieties, uncertainties, and aggressions which in their turn present other problems to the school. Most pupils soon claim the asylum of not being "ready" for arithmetic at a given level or, eventually, of not having a mathematical mind. Such explanations are readily seized upon by defensive teachers and parents. Few pupils ever reach the stage at which automatic reinforcements follow as the natural consequences of mathematical behavior. On the contrary, the figures and symbols of mathematics have become standard emotional stimuli. The glimpse of a column of figures, not to say an algebraic symbol or an integral sign, is likely to set off — not mathematical behavior — but a reaction of anxiety, guilt, or fear.

The teacher is usually no happier about this than the pupil. Denied the opportunity to control via the birch rod, quite at sea as to the mode of operation of the few techniques at her disposal, she spends as little time as possible on drill subjects and eagerly subscribes to philosophies of education which emphasize material of greater inherent interest. A confession of weakness is her extraordinary concern lest the child be taught something unnecessary. The repertoire to be imparted is carefully reduced to an essential minimum. In the field of spelling, for example, a great deal of time and energy has gone into discovering just those words which the young child is going to use, as if it were a crime to waste one's educational power in teaching an unnecessary word. Eventually, weakness of technique emerges in the disguise of a reformulation of the aims of education. Skills are minimized in favor of vague achievements — educating for democracy, educating the whole child, educating for life, and so on. And there the matter ends; for, unfortunately, these philosophies do not in turn suggest improvements in techniques. They offer little or no help in the design of better classroom practices.

There would be no point in urging these objections if improvement were impossible. But the advances which have recently been made in our control of the learning process suggest a thorough revision of classroom practices and, fortunately, they tell us how the revision can be brought about. This is not, of course, the first time that the results of an experimental science have been brought to bear upon the practical problems of education. The modern classroom does not, however, offer much evidence that research in the field of learning has been respected or used. This condition is no doubt partly due to the limitations of earlier research. But it has been encouraged by a too hasty conclusion that the laboratory study of learning is inherently limited because it cannot take into account the realities of the classroom. In the light of our increasing knowledge of the learning process we should, instead, insist upon dealing with those realities and forcing a substantial change in them. Education is perhaps the most important branch of scientific technology. It deeply affects the lives of all of us. We can no longer allow the exigencies of a practical situation to suppress the tremendous improvements which are within reach. The practical situation must be changed.

There are certain questions which have to be answered in turning to the study of any new organism. What behavior is to be set up? What reinforcers are at hand? What responses are available in embarking upon a program of progressive approximation which will lead to the final form of the behavior? How can reinforcements be most efficiently scheduled to maintain the behavior in strength? These questions are all relevant in considering the problem of the child in the lower grades.

In the first place, what reinforcements are available? What does the school have in its possession which will reinforce a child? We may look first to the material to be learned, for it is possible that this will provide considerable automatic reinforcement. Children play for hours with mechanical toys, paints, scissors and paper, noise-makers, puzzles — in short, with almost anything which feeds back significant changes in the environment and is reasonably free of aversive properties. The sheer control of nature is itself reinforcing. This effect is not evident in the modern school because it is masked by the emotional responses generated by aversive control. It is true that automatic reinforcement from the manipulation of the environment is probably only a mild reinforcer and may need to be carefully husbanded, but one of the most striking principles to emerge from recent research is that the *net* amount of reinforcement is of little significance. A very slight reinforcement may be tremendously effective in controlling behavior if it is wisely used.

If the natural reinforcement inherent in the subject matter is not enough, other reinforcers must be employed. Even in school the child is occasionally permitted to do "what he wants to do," and access to reinforcements of many sorts may be made contingent upon the more immediate consequences of the behavior to be established. Those who advocate competition as a useful social motive may wish to use the reinforcements which follow from excelling others, although there is the difficulty that in this case the reinforcement of one child is necessarily aversive to another. Next in order we might place the good will and affection of the teacher, and only when that has failed need we turn to the use of aversive stimulation.

In the second place, how are these reinforcements to be made contingent upon the desired behavior? There are two considerations here — the gradual elaboration of extremely complex patterns of behavior and the maintenance of the behavior in strength at each stage. The whole process of becoming competent in any field must be divided into a very large number of very small steps, and reinforcement must be contingent upon the accomplishment of each step. This solution to the problem of creating a complex repertoire of behavior also solves the problem of maintaining the behavior in strength. We could, of course, resort to the techniques of scheduling already developed in the study of other organisms but in the present state of our knowledge of educational practices, scheduling appears to be most effectively arranged through the design of the material to be learned. By making each successive step as small as possible, the frequency of reinforcement can be raised to a maximum, while the possibly aversive consequences of being wrong are reduced to a minimum. Other ways of designing material would yield other programs of reinforcement. Any supplementary reinforcement would probably have to be scheduled in the more traditional way.

These requirements are not excessive, but they are probably incompatible with the current realities of the classroom. In the experimental study of learning it has been found that the contingencies of reinforcement which are most efficient in controlling the organism cannot be arranged through the personal mediation of the experimenter. An organism is affected by subtle details of contingencies which are beyond the capacity of the human organism to arrange. Mechanical and electrical devices must be used. Mechanical help is also demanded by the sheer number of contingencies which may be used efficiently in a single experimental session. We have recorded many millions of responses from a single organism during thousands of experimental hours. Personal arrangement of the contingencies and personal observation of the results are quite unthinkable. Now, the human organism is, if anything, more sensitive to precise contingencies than the other organisms we have studied. We have every reason to expect, therefore, that the most effective control of human learning will require instrumental aid. The simple fact is that, as a mere reinforcing mechanism, the teacher is out of date. This would be true even if a single teacher devoted all her time to a single child, but her inadequacy is multiplied many-fold when she must serve as a reinforcing device to many children at once. If the teacher is to take advantage of recent advances in the study of learning, she must have the help of mechanical devices.

The technical problem of providing the necessary instrumental aid is not particularly difficult. There are many ways in which the necessary contingencies may be arranged, either mechanically or electrically. An inexpensive device which solves most of the principal problems has already been constructed. It is still in the experimental stage, but a description will suggest the kind of instrument which seems to be required. The device consists of a small box about the size of a small record player. On the top surface is a window through which a question or problem printed on a paper tape may be seen. The child answers the question by moving one or more sliders upon which the digits 0 through 9 are printed. The answer appears in square holes punched in the paper upon which the question is printed. When the answer has been set, the child turns a knob. The operation is as simple as adjusting a television set. If the answer is right, the knob turns freely and can be made to ring a bell or provide some other conditioned reinforcement. If the answer is wrong, the knob will not turn. A counter may be added to tally wrong answers. The knob must then be reversed slightly and a second attempt at a right answer made. (Unlike the flash-card, the device reports a wrong answer without giving the right answer.) When the answer is right, a further turn of the knob engages a clutch which moves the next problem into place in the window. This movement cannot be completed, however, until the sliders have been returned to zero.

The important features of the device are these: Reinforcement for the right answer is immediate. The mere manipulation of the device will probably be reinforcing enough to keep the average pupil at work for a suitable period each day, provided traces of earlier aversive control can be wiped out. A teacher may supervise an entire class at work on such devices at the same time, yet each child may progress at his own rate, completing as many problems as possible within the class period. If forced to be away from school, he may return to pick up where he left off. The gifted child will advance rapidly, but can be kept from getting too far ahead either by being excused from arithmetic for a time or by being given special sets of problems which take him into some of the interesting bypaths of mathematics.

The device makes it possible to present carefully designed material in which one problem can depend upon the answer to the preceding and where, therefore, the most efficient progress to an eventually complex repertoire can be made. Provision has been made for recording the commonist mistakes so that the tapes can be modified as experience dictates. Additional steps can be inserted where pupils tend to have trouble, and ultimately the material will reach a point at which the answers of the average child will almost always be right.

If the material itself proves not to be sufficiently reinforcing, other reinforcers in the possession of the teacher or school may be made contingent upon the operation of the device or upon progress through a series of problems. Supplemental reinforcement would not sacrifice the advantages gained from immediate reinforcement and from the possibility of constructing an optimal series of steps which approach the complex repertoire of mathematical behavior most efficiently.

A similar device in which the sliders carry the letters of the alphabet has been designed to teach spelling. In addition to the advantages which can be gained from precise reinforcement and careful programming, the device will teach reading at the same time. It can also be used to establish the large and important repertoire of verbal relationships encountered in logic and science. In short, it can teach verbal thinking. As to content instruction, the device can be operated as a multiple-choice self-rater.

Some objections to the use of such devices in the classroom can easily be foreseen. The cry will be raised that the child is being treated as a mere animal and that an essentially human intellectual achievement is being analyzed in unduly mechanistic terms. Mathematical behavior is usually regarded, not as a repertoire of responses involving numbers and numerical operations, but as evidences of mathematical ability or the exercise of the power of reason. It is true that the techniques which are emerging from the experimental study of learning are not designed to "develop the mind" or to further some vague "understanding" of mathematical relationships. They are designed, on the contrary, to establish the very behaviors which are taken to be the evidences of such mental states or processes. This is only a special case of the general change which is under way in the interpretation of human affairs. An advancing science continues to offer more and more convincing alternatives to traditional formulations. The behavior in terms of which human thinking must eventually be defined is worth treating in its own right as the substantial goal of education.

Of course the teacher has a more important function than to say right or wrong. The changes proposed would free her for the effective exercise of that function. Marking a set of papers in arithmetic — "Yes, nine and six *are* fifteen; no, nine and seven *are not* eighteen" — is beneath the dignity of any intelligent individual. There is more important work to be done—in which the teacher's relations to the pupil cannot be duplicated by a mechanical device. Instrumental help would merely improve these relations. One might say that the main trouble with education in the lower grades today is that the child is obviously not competent and *knows it* and that the teacher is unable to do anything about it and *knows that too*. If the advances which have recently been made in our control of behavior can give the child a genuine competence in reading, writing, spelling, and arithmetic, then the teacher may begin to function, not in lieu of a cheap machine, but through intellectual, cultural, and emotional contacts of that distinctive sort which testify to her status as a human being.

Another possible objection is that mechanized instruction will mean technological unemployment. We need not worry about this until there are enough teachers to go around and until the hours and energy demanded of the teacher are comparable to those in other fields of employment. Mechanical devices will eliminate the more tiresome labors of the teacher but they will not necessarily shorten the time during which she remains in contact with the pupil.

A more practical objection: Can we afford to mechanize our schools? The answer is clearly yes. The device I have just described could be produced as cheaply as a small radio or phonograph. There would need to be far fewer devices than pupils, for they could be used in rotation. But even if we suppose that the instrument eventually found to be most effective would cost several hundred dollars and that large numbers of them would be required, our economy should be able to stand the strain. Once we have accepted the possibility and the necessity of mechanical help in the classroom, the economic problem can easily be surmounted. There is no reason why the school room should be any less mechanized than, for example, the kitchen. A country which annually produces millions of refrigerators, dish-washers, automatic washing-machines, automatic clothes-driers, and automatic garbage disposers can certainly afford the equipment necessary to educate its citizens to high standards of competence in the most effective way.

There is a simple job to be done. The task can be stated in concrete terms. The necessary techniques are known. The equipment needed can easily be provided. Nothing stands in the way but cultural inertia. But what is more characteristic of America than an unwillingness to accept the traditional as inevitable? We are on the threshold of an exciting and revolutionary period, in which the scientific study of man will be put to work in man's best interests. Education must play its part. It must accept the fact that a sweeping revision of educational practices is possible and inevitable. When it has done this, we may look forward with confidence to a school system which is aware of the nature of its tasks, secure in its methods, and generously supported by the informed and effective citizens whom education itself will create.

Skinnerian Learning in Mathematics

Skinner, who has been responsible for the basic concepts behind programmed learning and teaching machines, distinguishes between two types of learning. He begins by noting that responses can be separated into two basic classes. The first of these are responses which follow an observable stimulus. Such elicited responses are classified as *respondents*. The blink of an eyelid when a bright light is flashed on the face would be an example of a respondent. We can condition a person to give this same response to a less basic stimulus — say, a loud noise. This conditioning is done like **Pavlov's** classic experiment. The flash of light and the loud noise are presented together again and again until the subject will blink when the loud noise is presented alone. The most crucial features in the conditioning process are the simultaneity of the two stimuli and the amount of repetition. Respondent or reflex learning pairs a response with a specific stimulus. What is acquired by such learning is a variety of reflexes — but certainly not mathematics nor the subject matter of most teachers.

As teachers, we must turn to Skinner's second category of responses to find classroomlike behavior. This category is comprised of responses which occur without any particular stimulus at all. These emitted responses he calls *operants*. Psychologists before Skinner recognized spontaneous or random responses, but they believed that such responses were caused by some unknown or unidentifiable stimulus. Skinner finds this explanation too "mysterious." He believes that operants simply occur, and that the stimulus conditions (if any exist) are irrelevant to the use and understanding of operant behavior. That the operant be reinforced is important. The law of operant conditioning states that *if the occurrence of an operant is followed by the presence of a reinforcing stimulus, the strength (or probability of that operant occurring again) is increased.*

There are two important items to note here. The first is that Skinner is most concerned with the stimulus which *follows* the response behavior. That is, what the teacher does to get the student to respond is not as important as the reaction or stimulus which the student receives from the teacher after he responds. This reaction shapes the chances of the student giving this operant response again or of his giving a similar response in the same class of responses. The second point to notice is that Skinner carefully avoids the invention of any intermediate connections such as "bonds" between stimulus and response. In fact, he insists that we must be careful to limit our theory to only observable behaviors and is, therefore, opposed to the use of such terms as "drive," "instinct," and "will power."

Skinner realizes that most human behavior involves operant responses. Therefore, operant learning is much more important than respondant learning. The initial emitted response or operant does not need to be exactly of the form or degree of the one we wish to teach. Operant responses are emitted with a range of form and intensity. If only the extreme values are reinforced, this range shifts to higher and higher values. By selective reinforcement, the initial operant response can be changed in a series of approximations until it approaches the final behavior that is desired. This successive approximation approach is obviously quite different from the conditioning or stamping-in of the Thorndike connections; it has come to be called *operant shaping*. While repetition has a place in increasing response strength, the mechanical repetition that conditioning brings to mind does not help to change responses in desired directions as the successive approximations of shaping do. Shaping also allows one to teach the subject how to do something that he has never done before, that is, to make a distinctive or novel response.

Reinforcing Responses

Responses are the most important aspect of operant learning, and the way they are reinforced determines most of the qualities of that learning. As might be expected, much of the work of Skinner and his colleagues has been to study the effect of different patterns and schedules of reinforcement. Most of these studies have been done with laboratory animals like rats or pigeons. Nevertheless, the results almost always have their counterparts in observable human behavior.

One of the first discoveries that Skinner made was that operants can be shaped without rewarding or reinforcing *every* response. If a pigeon is rewarded with a food pellet for pecking at a certain key, he will continue to peck at that key for several more times even though he is not rewarded with more food. It is not necessary to reinforce after every desired response but only intermittently during the course of several such responses. This observation led Skinner to study two basic patterns of reinforcement. In the first, interval reinforcement, a reward is given on a fixed interval of time — say, every three minutes. In the second, ratio reinforcement, a reward is given after a fixed ratio of responses — say, after every ten or fifteen responses have occurred. These numbers cannot be too large at the beginning of training, but they can gradually be increased to the order of one reinforcement to every 200 responses. Oddly enough, Skinner has found that the less frequent the reinforcement on a ratio schedule, the more rapid the response. That is, the animal behaves as if he knows that the faster he responds the faster he will be reinforced. Both fixed interval and fixed ratio reinforcement schedules are characterized by a pause in response just after a reinforcement. Animals seem to "know" that the responses made just after a reinforcement will never result in another immediate reinforcement. These pauses do not occur if the reinforcement schedule is made random. If the time interval size is varied at random or if the ratio interval size is varied at random, there is always a chance that the next response after reinforcement could result in another reinforcement, and the animal does not pause. (2:24)

Is this strange animal behavior reflected in human behavior? Indeed it is! Factory workers who are paid for every five units completed usually pause briefly after every money reinforcement before resuming their response rate. On the other hand, Las Vegas slot machine players working under an "occasional payoff" variable ratio schedule play machines without pause. Could you observe similar behavior in the classroom? A student who was to be graded after every ten homework problems should pause for some time before starting in on a new group of ten. In contrast, students who expected to be graded at random for the work they had completed (perhaps as the teacher walked around the room) should work without pause. Does this explain why many students "goof off" at the end of a mathematics class when time is given to work on a new assignment? Could you design an experiment to find out?

Forgetting and Extinction

When operant conditioned responses disappear or undergo *extinction*, the animal or human has forgotten. Skinner has shown that with animals, forgetting is not a function of time alone. Rather, animal forgetting comes about when reinforcement fails to follow the conditioned response. The animal continues to make responses, but at a decreasing rate until the responses cease entirely. The rate of decrease can be taken as a kind of measure of resistance to extinction — the smaller the rate of decrease the greater the resistance to extinction.

This resistance to extinction of a conditioned operant's response depends heavily upon the schedule of reinforcement that was used during the conditioning process. Paradoxically, conditioned responses which were developed under an intermittent reinforcement schedule are far stronger in resisting extinction than responses which were developed under continuous reinforcement. It is true that, in general, the more reinforcements that are given the greater the resistance to extinction. But if the total number of reinforcements is comparable, then intermittent conditioning is much superior.

Although forgetting in human beings is far more complex than extinction in animals, this advantage of intermittent conditioning explains very nicely some paradoxical human behavior. For example, it explains a man's persistent attempts to light a cigarette lighter which only worked occasionally in the first place. If the lighter had previously worked *every* time (continuous reinforcement), he would give up much more quickly. Similarly, the superiority of intermittent conditioning explains why children are so persistent in throwing temper tantrums even though they only *occasionally* influence mothers in the desired way. Can we see similar behavior in the mathematics classroom? It is not uncommon to find a class of children (often low achievers) who respond to a teacher's question about the answer to a mathematics problem with a series of rapid and frustratingly random guesses.

"If four times 31 is 124, what would eight times 31 be?"
"824?" "128?" "284?" "248?" "482?"
"Yes! The answer is 248."

What the teacher has done with this class does not have anything to do with patterns in multiplication. Rather, he has probably started to condition a whole process of random guessing as an acceptable procedure in mathematics. If the same procedure is intermittently reinforced, extinguishing this unwanted behavior may be very difficult later on. (It is true that guessing is useful in mathematics, but it is a plausible guessing that is valuable — not the random guessing prevalent in many students.)

Rewards and Punishment

Sometimes we may want to deliberately extinguish a particular response. We commonly do this by punishing the animal or the child. Punishment is certainly not a reinforcer, since it does not increase the probability of a response. On the other hand, Skinner's work has shown that punishment is not an exact opposite of a reinforcer either. When an animal is punished, its rate of responding is decreased but usually reappears sometime later. The total effect of punishment is temporary and unstable. Punishment seems to work by causing emotional agitation and withdrawal from the punishing situation. If these factors become generalized or dislocated, punishment may well produce surprising and unwanted effects. Suppose you beat your dog for chasing automobiles after he has returned home. The dog may become "emotional" not when chasing automobiles but when coming home, and it is the latter response which is apt to be suppressed. All too often punishment at school backfires in the same way. Children misbehave on the playground during recess. The teacher's wrath is aroused, and he punishes them when they return to the classroom by assigning twice as many math problems for homework. What the children learn is not that they must control themselves on the playground, but that they hate mathematics problems. How much of the prevalent dislike for mathematics can be traced to its use as a punishment for something else?

In general, a particular response cannot be completely eliminated from an organism's behavior by the use of punishment alone. While punishment may be used to hold a particular response at a low strength, it must be continued indefinitely since it can never completely eliminate the response. But because punishment is followed by a period of suppressed response, it is possible to take advantage of this period in order to strengthen some other response by reinforcement. If the response which is reinforced is incompatible with the one that is punished, then the subject may in a sense be "taught" not to make the forbidden response.

Nevertheless, punishment seems to be a poor way to control and modify behavior. It is deceptive, since it can initially have dramatic effects that are, in fact, only temporary. Furthermore, it may produce emotional side effects that may later be more undesirable than the original behavior.

S-R and the Role of the Teacher

How much mathematics learning takes place not for the pleasure of knowing, but because of the fear of failing? We are only now beginning to realize the crucial importance of structuring each day's mathematics lesson so that *all* children can be successful. There appear to be two alternative ways to do this — either water down the content or completely change the traditional role of the teacher. Few of us would choose the former, but what does the latter entail? We traditionally "cover" the material in our textbooks, letting those who can learn, learn, and those who cannot fail. Traditionally we expect all children to learn in the same way, using the same materials, guided by one teacher who teaches the entire class simultaneously.

How does operant conditioning change this classroom picture? First, it focuses on the *individual.* Operant conditioning says little about group behavior. Instead, it shapes individual behavior by starting with an individual operant shaped by reinforcements to each individual student, which are supplied according to an individual's responses or groups of his responses. The teacher who believes that children learn through operant conditioning processes believes that each child must be carefully accounted for in the teaching process, which means appropriate individualization of instruction.

Secondly, operant conditioning focuses on *behavior.* The learner's responses are shaped. Shaping implies activity. In this sense, operant conditioning is learning by doing, which suggests that actions or behaviors should be the ultimate goals of our teaching. Our teaching objectives should be stated in behavioral terms. The teacher who believes that children learn through operant conditioning processes is not satisfied to teach only for an appreciation of mathematics. Appreciation, though important, is ultimately an elusive goal. To decide what school practices best yield appreciation of mathematics is impossible simply because to really tell when appreciation has been achieved is impossible. A far better objective might be to teach so that students would go to the library and check out a book on mathematics without being required to. Such behavior can be observed. Because of its emphasis on behavior, operant conditioning challenges us to think of mathematics (and all school subjects) not as a passive collection of musty facts to be covered, but as an active, growing, dynamic way of processing information.

Finally, operant conditioning calls for the teacher to be an architect and builder of individual behaviors, his blueprints specifying not only which conceptual components to teach but also when to teach them. These plans call for a detailed task analysis of the concepts and skills to be taught which identify small components and their relationship to each other. These components should then be arranged into a sequence so that the learner moves step by step through a series of progressive approximations to the final behavior which is required. Great flashes of insight and leaps of reasoning which characterize mathematical genius are not appropriate to teaching (unless these specific shortcuts are the final objective of the lesson).

Building sequences of small components into teaching strategies may sound very nice in the abstract, but what does building mean in practice? Are there principles to guide the construction of the sequences? Such principles are now being developed from experimental studies of the learning of mathematics. The next module consists of an article which describes a method for conducting experiments and suggests a method for analyzing content components and ordering them into a sequence that will maximize learning and proficiency in mathematics.

Learning and Proficiency in Mathematics

Robert M. Gagné

In this report I should like to tell you about some of the ideas and products which have resulted from a collaboration of experimental psychologists, mathematicians, and mathematics teachers. The project which I have directed at Princeton University has been involved in such a collaboration with the University of Maryland Mathematics Project, which, as you may know, has as its Director, Dr. John Mayor, and as Associate Director, Mrs. Helen Garstens.

What we have done together is to develop a method of conducting experimental studies of the learning of mathematics. We have used this method to investigate the action of several factors in the learning situation and to verify their effects on mathematics learning.

What is the method we have developed and used? How has it worked out in revealing the factors at work in learning? And what kinds of results does it lead to, with their implications? Perhaps it should be said, at the outset, that my description of this method is intended to stimulate thinking and discussion about the process of instruction as an investigable set of events. The major purpose here is to present a viewpoint about research and its relation to the process of instruction.

The Method

The primary method employed involved the use of what are called "programmed learning" materials. As readers of the literature on teaching machines know, such materials are designed to present information to the learner, and he is required to make a response to it by filling in a blank or answering a question [1].* Once he has done this, an answer frame is exposed which informs him of the correct response; he then proceeds to the next frame, and so on throughout the program.

In the studies being described, learning programs were devised by reproducing typed frames of information on index cards, or on half-size sheets of paper, and assembling them into booklets of convenient size having loose hinges which permitted the pages to be flipped over easily. The answers to the questions posed by the frames were printed on the back of the cards. Generally, students were instructed to turn back the card and read the frame again whenever they had made an incorrect

---

Reprinted from *The Mathematics Teacher*, vol. 56 (December, 1963): 620-626. © 1963 by the National Council of Teachers of Mathematics. Used by permission.

*Numerals in brackets refer to the References at the end of this article.

response. Answers were recorded by the students on specially prepared answer sheets numbered to correspond with the frame member.

Using such materials, studies were conducted both in the laboratory with individual students one at a time, and in classrooms, where groups of students responded to the learning programs each in his individual manner. The setup used with individual learners consisted of a visible card file mounted on a stand, in which cards containing the individual frames were inserted. The materials used with students in classrooms were looseleaf booklets and an answer sheet. The frames of the learning program proceeded in a step-by-step fashion, each requiring a response on the part of the student.

When school groups were used, provision had to be made for the fact that different students would finish a booklet at different times, not only because of individual differences in learning rate, but also because the learning program itself was presented in experimentally different versions. This situation was handled by having each student, upon completion of a booklet, turn his attention to other work (unrelated to the program) which had previously been assigned by the regular classroom teacher. In this manner it was possible to have all students finish all booklets.

Content of learning programs

The learning programs employed in these studies had various mathematical contents keyed to the level of mathematical sophistication of the students. Generally speaking, topics for the programs were designed to be unfamiliar to the students, but involved the assumption of particular previous knowledge in each case. In some instances, programmed learning materials were based upon a topic occurring at a particular point in a mathematics textbook and were introduced at the normal time in the instructional sequence. In other cases, the materials were concerned with topics which the students had not yet reached in their normal curriculum. Specifically, studies used materials on the following topics: (1) deriving formulas for the sum of terms in number series [2], (2) "solving" simple algebraic equations [3], and (3) deriving definitions for, and performing additions of, integers [4]. In studies currently underway, materials on tangents of angles and on basic nonmetric geometry are being used.

Measuring student performance

Following the completion of learning programs, a number of measures of performance were employed to reveal what the students had learned. Typically, one of these was a test of performance designed to measure proficiency in the class of tasks specifically covered in the learning program. In most instances, a test of transfer was also employed, in the attempt to determine the extent of generalizability of what had been learned. Such tests presented problems which were new to the student and which belonged to a class of tasks other than that included in the learning program. For example, a test of transfer given following a learning program on addition of integers included problems on addition of rational numbers. Finally, a third type of measure was a test of subordinate knowledges, which was intended to reveal whether a student could or could not perform correctly each of the types of tasks contained within the learning program, but subordinate to the final task for which the program was intended.

Variables in learning programs

The basic situation with which we were dealing, then, was one in which a learner interacted with the material on a printed page. It is of some interest to consider what kinds of variables may be at work when an instructional process is generated in this way. What functions are being performed by these printed statements? What do they do for the learner?

As described more completely elsewhere [3], the functions which these statements seemed to us to be performing included the following: (1) they may define for the learner the general form of performance expected at the end of each subtopic; (2) they define unfamiliar words and symbols; (3) they may require the learner to recall certain subordinate knowledges he has previously learned; and (4) they "guide" his thinking about the new task, while encouraging discovery. It may be noted that in an experiment, any or all of these functions of frames may be manipulated in order to test their effects on the learning outcomes. For example, one can give more or less guidance to thinking within a learning program. Or one can vary the number of times a newly acquired task is recalled and thus make repetition (in this specific sense) an experimental variable.

But there is another way in which one can vary the frames of a learning program, and this may be the most important of all. This is by ordering the topics throughout the program. By "topic" is meant a distinguishable principle of knowledge which can govern the performance of a class of human tasks. When we consider a final performance to be learned (such as "adding integers"), we find that it can be analyzed into a number of subordinate topics which must first be mastered before the final task can be attained. These topics in turn depend upon the mastery of other subordinate topics. This kind of analysis, then, results in the identification of a hierarchy of subordinate knowledges, such as that shown in Figure 17. Each element of this subordinate knowledge is hypothesized to support the learning of each topic in the hierarchy to which it is connected by an arrow. That is, mastery of the subordinate knowledge is considered to be essential to the attainment of a related higher-level topic; learning of the latter cannot occur without it. In Figure 17, for example, a task like IVa, "using 0 as the additive identity" must be mastered (according to hypothesis) in order for IIIa, "stating and using the definition of addition of an integer and its additive inverse," to be learned. Similarly, IIIa must first be mastered before IIa can be achieved, and so on.

The implications of this hypothesis about topical order are easy to grasp. If an individual learner has achieved the subordinate knowledge represented by IVa (Fig. 1), his learning of IIa will be highly probable if he has also learned IIIa, and highly improbable if he has not. When translated into a formula for the design and construction of a learning program, or other instructional sequences, this means that the topic IVa must precede the topic IIIa, which in turn must precede the topic IIa, and so on. If the order is violated, or an intervening topic omitted, the acquisition of knowledge of any superordinate topic will be unsuccessful.

The method of making an analysis to arrive at the set of subordinate knowledges arranged in a hierarchy like that of this figure should be briefly mentioned. One begins with the terminal class (or classes) of tasks for which learning is being undertaken. For each of these, one asks the question, "What must the learner already know how to do, in order to achieve this (new) performance, assuming that he is to be given only instructions?" The latter part of this question assumes that the instructions will have the functions of frames previously mentioned, with the exception of task repetition. The answer to this question defines one or more elements of subordinate knowledge (at level I in Fig. 17). The question is then applied to each of these in turn, thus identifying the entire hierarchy. The process ends when one arrives at subordinate knowledge (like Va and Vb) which can be assumed to be possessed by every learner for whom the learning program is intended.

This description should not be taken to imply, however, that there is only one unique hierarchy appropriate to the learning of any particular final performance. This particular one is based upon a logical sequence developed in UMMaP materials; had we used an SMSG chapter, the hierarchy would perhaps have been somewhat different. And, of course, it would be perfectly possible to construct a very simple hierarchy for Task 2 (adding integers) based upon three computational "rules." The importance of such hierarchies is not that they can be uniquely determined, but rather that they depict a learning structure which, once defined, indicates steps in instruction which must be accomplished in a proper sequence in order to achieve the desired performance.

The Method in Use

The approach described separates the factors which potentially influence mathematics learning into two broad categories. The first of these may be called *instructional* variables, conceived as a set of functions that are performed in presenting *each* new topic, or new principle, to be learned. As previously mentioned, they include the functions of definition of new stimuli and terminal performances, recalling, guiding thinking, and task repetition. The second broad category is *topical order*, which pertains to the sequence of subordinate knowledge (topics) that must be acquired in order to achieve some final performance. Any instructional sequence, one form of which is a learning program, utilizes certain selected values and arrangements of variables from each of these classes.

How do we actually go about putting this method into use in performing experimental studies?

First, we have to analyze a subject to be taught, to reveal its knowledge structure. By so doing, we are defining the subordinate knowledge that students must possess in order to master the task.

Second, we construct a basic learning program having the characteristics I suggested. It has a *topic order* determined by the subordinate-knowledge hierarchy. In proceeding from one subtopic to the next, care

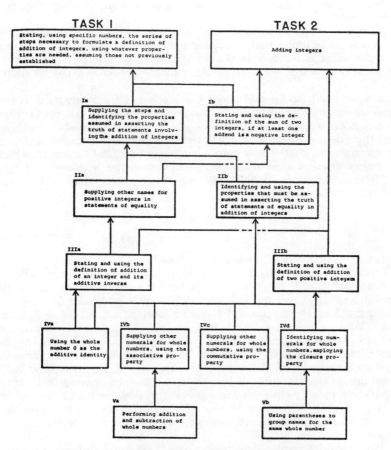

TASK I

Stating, using specific numbers, the series of steps necessary to formulate a definition of addition of integers, using whatever properties are needed, assuming those not previously established

TASK 2

Adding integers

Ia

Supplying the steps and identifying the properties assumed in asserting the truth of statements involving the addition of integers

Ib

Stating and using the definition of the sum of two integers, if at least one addend is a negative integer

IIa

Supplying other names for positive integers in statements of equality

IIb

Identifying and using the properties that must be assumed in asserting the truth of statements of equality in addition of integers

IIIa

Stating and using the definition of addition of an integer and its additive inverse

IIIb

Stating and using the definition of addition of two positive integers

IVa

Using the whole number 0 as the additive identity

IVb

Supplying other numerals for whole numbers, using the associative property

IVc

Supplying other numerals for whole numbers, using the commutative property

IVd

Identifying numerals for whole numbers, employing the closure property

Va

Performing addition and subtraction of whole numbers

Vb

Using parentheses to group names for the same whole number

Figure 17. A Hierarchy of a Learning Program.

is taken that individual frames have performed the functions previously mentioned, namely, the definition of terminal performances and new terms, the activation of recall, and the guidance of thinking.

Having done this, we next introduce the *variations* in this basic program which we are interested in from an experimental standpoint. For example, if we want to study the effects of more or less guidance to thinking, we construct programs containing more or fewer frames performing this function. Or, if we want to study the effects of task repetition (because of its presumed influence on recallability), we construct several forms of programs providing variation in this variable.

Next, we administer these programs to classes of school children. Usually, they are divided into booklets in such a way that everyone finishes on each day.

Following this, we measure performance on a test designed to measure exactly what the program as a whole was designed to teach, no more and no less. In addition, if we are interested in the question of transfer of training, we administer a test involving some related but entirely new subject matter.

As a final step, we measure each person's subordinate knowledge — each one of the topics that is represented in the figure previously shown. If the student is not able to do the final task to perfection, we want to know why — which element of the subordinate knowledge he didn't have.

A word needs to be said about the scoring of the items used to test whether the individual could or could not perform the task set for him. These items were scored independently by two mathematics teachers, and their degree of agreement was virtually perfect. Of course, it should be remembered that this is somewhat unconventional "testing," which makes no attempt to measure degrees of knowledge, whatever that may be. The performance of each task was either wholly right or was considered wholly wrong.

Some Results

Let me describe briefly some of our results.

First, what about this knowledge hierarchy, and the topical order it determines? It will be recalled from the example shown in Figure 10, that a topic (i.e., an item of knowledge) at a particular level in the hierarchy may be supported by one or more topics at the next lower level. What is being predicted, then, is this: An individual will not be able to learn a particular topic if he has failed to achieve *any* of the subordinate topics that support it. This hypothesis can of course be tested at a number of different points throughout the hierarchy. With reference to the structure of Figure 10, our results showed that *for no topic were there more than three percent of instances contrary to this hypothesis* [4]. (We have no way of accounting for the three percent except as measurement error; however, we consider it to be gratifyingly small.)

These results imply something about individual differences. What they imply is that the most important difference among learners in their ability to perform a final task resides in their possession, or lack of possession, of this subordinate knowledge. Patterns of subordinate knowledge exhibited by successful and unsuccessful learners differ in quite predictable ways [3]. The unsuccessful learner, in progressing through the self-instructional program, effectively stops learning somewhere along the road through the knowledge hierarchy, and is unable to master any subordinate knowledge beyond that point.

As for a general aptitude like "intelligence," there is not surprisingly a moderately low degree of correlation with *rate* of progression through a learning program, and this correlation remains relatively constant. This correlation with rate, however, is to be distinguished from one with proficiency. The important thing suggested by our results is this: If all learners are allowed time to complete the program (as was true in our administration), and if this program is otherwise effective, their performance at the end comes to be *independent* of ability scores measured before the learning began, and highly dependent upon the specific subordinate knowledge they have learned. For example, in the study of learning about addition of integers, which turned out to have a moderately successful program (as indicated by average scores of 8.9 in achievement of the component knowledge, out of a total of 12), no significant relationships were found between mathematics grades before the experiment began and performance on the final task [4]. But there is a high correlation (.88) between the latter performance and the scores indicating the number of elements of subordinate knowledge learned.

What about instructional variables? As I have pointed out, it is possible to perform studies in which various modifications of these variables are incorporated into different forms of learning programs. One of our studies attempted to learn about the effects of two of these variables which can be manipulated by changing some features of the instructions provided in the frames of learning programs. One of these was task repetition — the number of additional examples given to the learner after he had first achieved each task of the knowledge hierarchy. A form of learning program characterized by "low repetition" gave one or two additional (varied) examples, while one having "high" repetition provided four or five times this number of examples, for each topic. It should be emphasized that what were given repetition were these subordinate tasks (see Fig. 1 for their names) in their terminal form, and not the remaining frames of the program.

Another factor subjected to variation was amount of guidance, represented by the number of frames taken to "instruct" the learner how to go from one topic to the next. One form of program used only a few frames to suggest a line of thinking to the learner ("low" guidance), while another used two or three times as many ("high" guidance). Neither of these forms could be said to "state the answer"; both required discovery on the part of the learner.

The results obtained on the effects of these variables can be rather simply stated. Neither repetition alone, nor guidance alone, had significant effects on the learning as measured by tests of final performance. When "high" repetition and "high" guidance were combined, however, this learning program produced a significantly higher number of successful learners than did the opposite combination of "low" repetition and "low" guidance [4], while other comparisons were not significantly different.

The results of course do not indicate a particularly strong effect by either of these learning variables. In fact, when the results of several of our studies are considered together, one does not gain the impression that instructional variables (such as guidance and repetition) have very pronounced effects upon the learning that takes place within the framework of an instructional program. This sends us back to a consideration of the contrastingly prominent effects of what may be called "content" variables, pertaining to the structure and organization of the knowledge being taught.

The most prominent implication of the results of these studies to date is that *acquisition of new knowledge depends upon the recall of old knowledge*. When stated in this way, the proposition seems to have few startling characteristics. In more specific form, however, it means that the learning of any particular capability requires the retention of other particular items of subordinate knowledge — not just any knowledge, or knowledge in some general sense. The learner acquires a new item of knowledge largely because he is able to integrate previously acquired principles into new principles, and he cannot do this unless he really knows these previously learned principles. The design of an instructional situation is basically a matter of designing a *sequence of topics*.

Our results imply that there are many, many specific sets of "readiness to learn." If these are present, learning is at least highly probable. If they are absent, learning is impossible. So, if we wish to find out how learning takes place, we must address ourselves to these specific readinesses to learn." If these are present, learning is at least highly probable. need to know a lot more about how they get established, and why they sometimes do not. The arrangement of the external conditions for learning is a matter of careful organization of the entities of knowledge, and their presentation in such a manner that no learner can help acquiring the new capabilities for achievement that we want to give him.

References

1. Lumsdaine, A. A., and Glaser, R. *Teaching Machines and Programmed Learning.* Washington, D.C.: National Education Association, 1960.
2. Gagné, R. M. "The Acquisition of Knowledge," *Psychol. Rev.,* LXIX (1962), 355-365.
3. _____ and Paradise, N. E. "Abilities and Learning Sets in Knowledge Acquisition," *Psychol. Monogr.,* No. 518 (1961), 75.
4. _____, Mayor, J. R., Garstens, H. L., and Paradise, N. E. "Factors in Acquiring Knowledge of a Mathematical Task," *Psychol. Monogr.,* No. 526 (1962), 76.

Hierarchies and Lesson Planning

As we have seen, the reinforcement and shaping principles of stimulus-response learning theory suggests that content should be broken into small component steps which can be quickly reinforced. We must realize that such steps cannot be presented in isolation, however. What we want to teach in mathematics is not just a collection of ideas, but a sequence of *organized knowledge*. We have already seen that the ideas we call mathematics can be organized in two fundamental ways. They can be organized *logically* according to the relationships that we can demonstrate between different ideas utilizing necessity and implication. These same ideas can also be organized in a *psychological* sense according to the way they are developed and learned by children and adults. This psyhological organization should be the primary guide for the teacher in planning lessons for the classroom. In most cases, psychological and logical organizations are complementary, but where differences arise, the teacher must first use psychological organization to establish basic ideas and then review the materials in the logical organization to give students an understanding of the ideas as they are organized by mathematicians.

Piaget shows many differences between logical and psychological organization in mathematics. His work suggests that the child develops mathematical ideas which are parallel in many ways to the logical foundations of mathematics. Nevertheless, the development of these ideas in the course of a child's development often shows marked differences from the way a mathematical logician would prefer to extend them. If Piaget's work separates logical and psychological distinctions, Gagné's work reunites them. His analysis of psychological prerequisites rests upon the logical connectives that make one piece of mathematics subordinate knowledge to another.

Gagné's psychological requirements are not as surprising or unexpected as those of Piaget. In fact, Gagné's empirical conclusions may seem so obvious to a mathematician that they provide few suggestions for teaching. However, Gagné not only suggests planning, his work indicates a particular kind of planning. Too many teachers see planning as a day-to-day task. In looking at a large sequence of prerequisites, a Gagné hierarchy goes well beyond daily planning and into the realm of unit planning.

Constructing Hierarchies

Working through a hierarchy is definitely a two-way process. To *use* a hierarchy a teacher begins at the bottom with the most basic and fundamental ideas and moves toward the top to the ultimate tasks or topics. But in *constructing* a hierarchy the teacher works "from the top down." He begins by considering the final goal or objective. If he is making a hierarchy to use with a particular textbook chapter, he begins at the *end* of the chapter. He looks for the kind of task the authors expect the student to successfully perform at the conclusion of his work. For many chapters the requirement may actually be the identification of several distinct tasks which have been grouped together. Most texts contain a set of review exercises or sample test items at the end of every chapter. Analyzing these items is usually the easiest way to identify the tasks which make up the top level of the hierarchy. If no textbook is being followed (e.g., if the teacher is preparing a supplementary unit), these procedures suggest that the first step in developing materials might be to write out a set of sample test items. Undoubtedly, these items need to be refined and changed as work developing the unit progresses. Nevertheless, beginning with test items helps focus on final goals and objectives.

The next step is to move down one level in the hierarchy and determine what kinds of behaviors must be combined to perform each of the final tasks. When several final tasks are involved, working separately on a hierarchy for one task at a time is usually best. If the separate hierarchies repeat the same subordinate knowledges, this repetition usually suggests how to merge them into a single hierarchy. Such repetition is often not apparent until one has reached the lowest levels of the hierarchy, however. End-of-chapter test and review items are again helpful in determining the second level of a hierarchy. Most authors include test items which are specifically focused at subordinate skills as well as final tasks. Studying the section headings within the chapter is also helpful. One must not assume, however, that all the section headings represent level two skills. Some sections, particularly those toward the beginning of the chapter, develop knowledge which is to be used and applied in later sections. Such knowledge belongs, of course, in lower sections of the hierarchy. Lower sections can also be determined by looking at the organization of material within sections of the chapter. As before, looking at exercise problems which are included within the chapter helps identify the subordinate knowledges which the author expects students to acquire in the course of their study.

Identifying the last and most fundamental levels for a hierarchy may pose a slightly different problem. Often material for this level is not specifically contained within the chapter being analyzed. One has to decide what the authors expected the child to be able to do when beginning the chapter. Usually, author expectations must be inferred from a careful inspection of previous materials. Sometimes previous chapters may also be helpful, but they tend to present a wealth of irrelevant detail and should be approached cautiously. The problem of deciding where to end a particular hierarchy is not a trivial one. The lowest level of the hierarchy in module 3-3 stops with the ability to add and subtract whole numbers and the ability to use grouping parentheses. Obviously, entire hierarchies could be constructed for each of these abilities. In the case of the ability to add and subtract whole numbers, the hierarchy would be substantial! Continuation of these hierarchies would result in analysis of most of the entire school mathematics program. To avoid this extensive analysis, rather arbitrary decisions must be used for terminating hierarchies. Chapter divisions in textbooks form one basis for these decisions. Natural divisions in the school year schedule also help guide the decisions.

Obviously, constructing a hierarchy for a textbook chapter involves a lot of work. Why should the teacher bother? Why not just follow the textbook day-to-day? If textbook writers developed their materials from carefully constructed hierarchies and published those hierarchies so that they could be evaluated, relying on textbooks to do unit planning might be possible for the teacher. Unfortunately many textbooks are *not* constructed from hierarchies. As a result, if a teacher follows a mathematics textbook blindly, it can ultimately lead him right into trouble. Constructing a hierarchy for particular textbook chapters is the most reliable way of uncovering these trouble spots in advance. This trouble can be in the form of subtle prerequisites which the authors overlooked or incorrectly assumed that students would already have. The hierarchy may reveal some subordinate knowledge treated in the wrong sequence. Or the hierarchy may reveal that the textbook includes side excursions into material which is not necessary for the final task. When this happens, the teacher has two options. He may prefer to postpone treatment of this extra material. Instead of confusing students about the goal they are seeking, this material might be better treated later as an example of an application or different analysis of the task. Or the teacher may prefer to simply delete the nonessential material from his course. Mathematics teachers are almost always faced with the problem of what material to delete at the end of the school year. As a result, entire chapters and topics are dropped under the pressure of time. Dropping extraneous material in the first and middle sections of the year, so that important topics would not have to be rushed or dropped at the end of the year seems much more enlightened.

What hierarchies provide, ultimately, is a way for a teacher to take charge of his own course instead of yielding every decision to a faceless stranger. Constructing hierarchies is hard work, but the final payoff is more freedom for the classroom teacher. Once the structure of mathematical prerequisites is clear, the teacher will see new ways to arrange his course. Often entire chapters may be moved into a new sequence that will more quickly arrive at the interests of students. This kind of shuffling is very dangerous without a hierarchy, for the teacher may arrive four or five weeks later at a section which students are totally unprepared to do. Constructing a hierarchy ahead of time alerts the teacher to such potential problems. Fortunately, hierarchies for a given textbook do not need to be constructed year after year. Once the basic hierarchies have been made, the teacher can use these insights to introduce and evaluate *variations* in the basic program each year.

Using Hierarchies

Hierarchies are constructed from "top to bottom." To use a hierarchy, however, the teacher begins at the bottom with the most basic ideas and moves upwards towards the ultimate topic. For example, the following Gagné hierarchy was used in a study of the retention of topics in elementary nonmetric geometry at the junior high school level. (3:123-131)

Ideas of similar complexity and difficulty are grouped together at a common level in this hierarchy. Level VI contains the simplest and most basic ideas — the ideas of separation and point. Level V is a slightly more complex idea — that a line consists of a set of points. From these primitive beginnings, the student then learns to draw and identify a straight line (level IV), a line segment and a ray (level III).

Notice that principles IIIa, IIIb, and IIIc each depend directly upon principles IVa and VIa. The hierarchy indicates that principles IIIa, IIIb, and IIIc cannot be learned until principles IVa

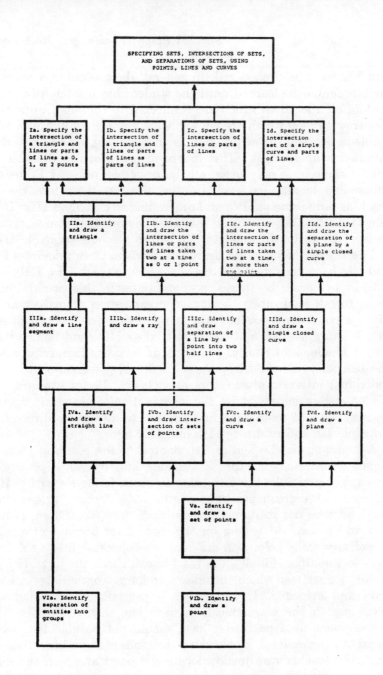

Figure 18. A Hierarchy of Principles.

A hierarchy of principles (definitions) to be acquired in a topic of elementary geometry. The topic to be learned is shown in the topmost box. (R. M. Gagné and O. C. Bassler, "Study of Retention of Some Topics of Elementary Nonmetric Geometry," J. Educ. Psychol. 54 (1963): 123-131.)

and VIa have been mastered. In general, the content in a particular box cannot be learned until the student has mastered the content in *all* the boxes which are connected to the box in question by arrows. Boxes which are not connected by arrows indicate independent content. Principle IIIa is independent of all the other knowledge on level IV. This independence provides the teacher with a choice. He may teach all the principles on level IV before proceeding to another level. But he also has the alternative of teaching only principle IVa and proceeding to principles IIIa, IIIb, and IIIc. If he chooses to do the latter, he must, of course, return to level IV to teach principle IVc before teaching principle IIId.

The teacher has another degree of freedom. Once principles IVa and VIa have been met, the teacher is free to teach IIIa, IIIb, and IIIc in any order he thinks is appropriate. In this case, he may want to fall back upon the *logical* organization of mathematics. Since a point divides a line into two half lines, each of which is a ray, he may wish to first teach IIIc, then IIIb, and finally IIIa. These freedoms of choice, which occur in every hierarchy, allow the teacher to alter flexibly the sequence of presentation to meet individual interests of students and classes. If, for example, his students seem impatient for the introduction of geometric figures, he may wish to teach IIIa first so that he can move directly to principle IIa and return to IIIb and IIIc later.

A hierarchy is checked and evaluated in the teaching process. Gagné's work suggests that if students fail to learn a particular concept or principle, the first thing to check is the hierarchy. Has an essential prerequisite been overlooked? Perhaps proper connections were not made between subordinate skills. Often, principles which seem to belong on the same level because of similar complexity really belong on different levels because one is subordinate to the other. Finally, the teacher can check the hierarchy by giving a final test which includes test items appropriate to each level and principle. If knowledge is psychologically structured according to the hierarchical organization, then a student who misses items at a particular level should not be able to answer items at a higher level. By checking the kinds of items that students miss, the teacher can quickly locate the point at which the difficulty occurred.

Daily Planning

The large scale unit planning reflected in a content hierarchy is useful for daily lessons, but it does not supply all the details necessary for a daily lesson. Planning the daily lesson should begin by identifying the three or four principles that are to be taught in the day's class. Studying the hierarchy, as well as the spirit of the underlying mathematics, should help the teacher decide the order in which these principles should be presented. Principles or content should be stated in terms of goals or objectives, and these objectives should specify the kind and amount of behavior expected from the pupil.

The next step in planning is to determine *how* the content of the lesson is to be presented. At this point, many teachers make a serious mistake. They concentrate on what *they* will do, what questions *they* will ask, what examples *they* will give. The mistake with this process is that it virtually ignores the student. Planning how the lesson will be conducted by focusing on what the *student* will do is more profitable. When the teacher plans in terms of his own actions, the result is often that the student is left with nothing to do except be a passive receiver of information. Shifting planning to the realm of student actions avoids this difficulty. It is also in line with operant shaping which holds *response* to be the key in learning.

In particular, the teacher's planning should center on providing students with a variety of actions in a single class period. Perhaps they can make responses by manipulating physical materials, as with compass-and-straightedge constructions or mathematics laboratory materials. Perhaps they can respond by answering questions, or by a string of responses. Breaking the class into small groups and letting students explain relationships to one of those groups might encourage these responses. If we need to strengthen specific responses, then having students work problems in class where they can receive needed reinforcement would be appropriate. The list of possibilities could go on. A variety of activities which not only change from day to day, but from section to section of a single lesson must be provided. By focusing upon students in the process of planning the lesson, the teacher is encouraged to seek this variety.

The next step in planning is to select the materials to be used. Obviously, this selection includes any physical equipment needed for the lesson. But one should also consider mathematical examples and problems as materials. Examples and problems which make important points about the content being taught should be chosen. Different examples should point up the importance of each different variable in the situation. In addition, alternate examples and problems which use the different modes of explanation discussed in Unit One should be selected.

The last step is to plan the opening and conclusion of the lesson. These sections deserve special emphasis because they provide the continuity between lessons. A strong opening and conclusion make it clear that mathematics is an organized body of knowledge. They relate the day's lesson to the preceding lesson and to subsequent lessons. Without this relationship, the responses taught in mathematics can quickly become a bag of unrelated tricks. Achieving continuity requires more than saying, "As you remember from last time . . ." Actually repeating some of the activities done "last time," and giving problems or exercises that will be useful "next time" achieve continuity.

Daily planning, then, requires that these items be determined:
    Content Objectives
    Action
    Materials
    Continuity
How much detail should be taken into account when planning? The answer to this question depends upon the needs of each individual teacher. Nevertheless, there is one negative guideline that can be followed. Lesson plans should not be written out in such detail that they force the teacher to be rigid. The teacher should always be ready to take advantage of unexpected student responses. If hours have been spent writing out the lesson plan, too much time will be invested to allow it to be changed. On the other hand, insufficient planning may prevent changes, too. This problem is especially true in the areas of Materials and Continuity. New teachers should overplan in these areas by selecting examples and problems to meet a wide variety of possible student responses. Having extra examples and problems available also avoids running out of things to do—a fairly common occurrence with new teachers who are uncertain about their timing.

For Further Investigation and Discussion

1. In investigation 3-1, we defined meaning in terms of the number of associations that could be made with a word in a given time period. We then *assumed* that words made of nonsense syllables were low in meaning. Was this assumption warranted? What kinds of associations might be made with a word like "zuhap"? Find a subject and have him write down free associations over a two minute time span for each word on our eight word list. Do the results suggest an alternate interpretation for your experimental results?

2. A standard reference for the study of the role of meaning in learning is Katona George, *Organizing and Memorizing.* New York: Columbia University Press, 1940. Katona describes many experiments which distinguish between meaningful and nonmeaningful learning. Choose any one of these and write a brief report and critique.

3. Read Skinner, B. F. "Teaching Science in High School—What is Wrong." Paper presented at a meeting of the American Association for the Advancement of Science, December, 1967. (Available from ERIC Document Reproduction Service as document ED 020 140.) Skinner discusses shortcomings of investigation and inquiry as they are used in the classroom. Although most of the references are to science teaching, many of them have counterparts in mathematics teaching. In addition, you will find some surprising criticisms of programmed learning and computer assisted instruction. Write a short paper explaining how Skinner's criticisms of science teaching might be applied to mathematics teaching. If you disagree with him, indicate why.

4. Stimulus-response learning is criticized because it gives no importance to structural relationships between concepts or the different responses to be learned. How does the work of Gagné help to fill in this oversight? In what ways does it still fail to account for meaning and structure in the learning process?

5. Successful operant conditioning depends upon the specification of behaviors to be obtained. The ability of a teacher to write behavioral objectives is crucial. Read the short book by Mager, Robert F. *Preparing Instructional Objectives.* Palo Alto, Calif.: Fearon Publishers, 1962. After taking the self-test in the book, write five objectives for a chapter in a mathematics textbook (of your own choosing).

For Lesson Planning

Analyze a chapter from a mathematics textbook into a hierarchy of ideas. Remember that the sequence in which principles are presented in the chapter is not the same as the hierarchy because of the freedom of choice that we have discussed both within and between levels. You should begin by reading the chapter and identifying the final goals of the author(s). Then work backwards to identify the subordinate principles which a student should know before he learns about these goals.

Compare the hierarchy you developed with those done by your classmates. How do they differ? Which differences are critical, and which merely reflect freedom of choice? What could you do to resolve these critical differences? Suppose you had analyzed the same topic as it was treated by a different textbook. Would you obtain the same hierarchy? Why or why not?

For Microteaching

Reinforcement schedules of operant conditioning suggest that one of the most important aspects of teaching is the teacher's response to pupil statements or questions. Microteaching provides a good way to study your own responses to students. In order to study your responses, you will have to plan a short lesson which encourages lots of student response. Plan your entire microlesson around a series of questions. For example, you might want to teach some of the tests for divisibility. You could begin the lesson by asking, "Can anyone tell me if 14,796 is divisible by 2?" "Did you have to do all the division before you could tell whether or not there would be a remainder?" "Is there a special place to look that will tell you if a number is divisible by 2? "Do you ever have to look at more than the last digit?" "Why not?" "Can you tell if 14,796 is divisible by 3?"

Although your microteach should not last more than four or five minutes, you should plan lots of questions in advance. Unless you know your students very well, this microteach will be a very difficult one to time in advance. If the subject matter is new, you may not use more than three or four of your questions. But if students know about your topic, they will give correct answers very quickly. Therefore, it is a good idea to overplan this lesson so that you are prepared for the unexpected!

Some people have difficulty getting students to respond, often because they are impatient. They ask a question, and when no answer is forthcoming, they supply an answer themselves. Students begin to realize that they never have to respond — the teacher will always provide the answer. If you have this difficulty, you need to practice keeping your mouth shut! After you ask the question, be silent. Eventually someone will say something, even if it is only "what do you mean?" One of the occupational hazards of teaching is the tendency to talk too much. Make silence work *for* you by encouraging more student response.

Another common difficulty in questioning is a tendency to talk to a single student and ignore the rest of the class. This tendency is especially easy when responding to a student-initiated question. One way to overcome this problem is to simply repeat the student's question, giving someone else in the class a chance to answer. One student's problem becomes a concern to the entire class. If you are dragged off into a side discussion because of a student's answer, try turning the answer into a question by asking the rest of the class if the same thing can be said in a different way.

Both the use of silence and the turning of students' questions to the rest of the class can be reinforcers just as strong as words of praise or pats on the head. They tell the students that you care about what they think, that their questions are important enough to ponder.

When you have planned your lesson, teach it to a class of five or six students. Record your teach on videotape or audiotape. As you teach, your immediate objective should be to encourage a maximum amount of student response.

When you have finished, view your videotape (or listen to the audiotape). On the first time through, evaluate your questioning style. Are your questions clear and brief? Do you make use of student ideas and responses to generate new questions? Do you listen carefully to what students say?

Now view or listen to the tape a second time. This time make a note of which students respond during the lesson. Have you distributed our questions among the class members? Did you encourage reticent students to contribute, or did you let one or two students dominate the discussion?

View or listen to the tape one last time. This time focus on your responses to students. Are you reinforcing the kind of behaviors you want? Have you unintentionally encouraged or reinforced behaviors you did not want? Were any of your responses intimidating? Could students be afraid to offer a suggested answer for fear it might be wrong?

If possible, improve and reteach the lesson to a different group of students. This particular activity requires you to do a lot of thinking on your feet, which is a skill that comes only with practice.

For Related Research

Two areas of research in mathematics education are related to this unit. The first of these is the importance and effect of reinforcements in the mathematics classroom. A comparison of the arithmetic achievement of fifth grade students who were always given immediate knowledge of test results with that of students who were always told results after a twenty-four hour delay is reported in Hillman, Bill W. "The Effect of Knowledge of Results and Token Reinforcement on the Arithmetic Achievement of Elementary School Children." *The Arithmetic Teacher* 17 (1970): 676-682. Results favored the children who received immediate feedback.

A similar study is reported in Paige, Donald D. "Learning While Testing." *Journal of Educational Research* 59 (1966): 276-277. An earlier study in this area is Brown, Francis J. "Knowledge of Results as an Incentive in School Room Practice." *Journal of Educational Psychology* 23 (1932): 532-552.

A study in which reinforcement was given by means of plastic tokens which could be exchanged for toys or candy is reported in Heitzman, Andrew J. "Effects of a Token Reinforcement System on the Reading and Arithmetic Skills Learnings of Migrant Primary School Pupils." *Journal of Educational Research* 63 (1970): 455-458.

The giving of tokens is a direct form of reinforcement. Less basic reinforcers, such as a word of encouragement or even a smile, are known as secondary reinforcers. A study using these more subtle reinforcers is Doherty, Anne and Richard A. Wunderlich. "Effect of Secondary Reinforcement Schedules On Performance of Problem-Solving Tasks." *Journal of Experimental Psychology* 77 (1968): 105-108. Another study involving reinforcement schedules is McNeil, John D. "Prompting vs. Intermittent Confirmation in the Learning of a Mathematical Task." *The Arithmetic Teacher* 12 (1965): 533-536.

Another major area of research is the development of hierarchies of prerequisite knowledge for mathematical topics. For more detailed descriptions of Gagné's experiments with hierarchies see Gagné, R. M. "The Acquisition of Knowledge." *Psychological Review* 69 (1962): 355-365. See also Gagné, R. M., J. R. Mayor, H. L.

Garstens, and N. E. Paradise. "Factors in Acquiring Knowledge of a Mathematical Task." *Psychological Monographs* 76 (1962).

References to Unit 3

1. Brownell, William A. and Harold E. Moser. "Meaningful vs. Mechanical Learning: A Study in Grade III Subtraction." *Duke University Studies in Education* 8 (1949): 1-207.
2. Cohen, Jozef. *Operant Behavior and Operant Conditioning.* Chicago: Rand McNally & Co., 1969.
3. Gagné, R. M. and O. C. Bassler. "Study of Retention of Some Topics of Elementary Nonmetric Geometry." *Journal of Educational Psychology* 54 (1963): 123-131.
4. Skinner, B. F. "The Science of Learning and the Art of Teaching." *Harvard Educational Review* 24 (1954): 86-97.

Unit 4

Gestalt Learning and
Heuristic Teaching

Introduction

In the process of analyzing the data of module 3-1 you plotted the number of correct responses in each trial against the number of that trial. Such a plot is a display of the progress made during the learning task as a function of time (or trial). The shape of this learning curve can also provide clues as to whether we should accept or modify S-R learning theories.

Consider the solution of a problem according to stimulus-response theories. The learner makes a response to the problem (stimulus). This response is shaped according to the results it produces. If the results are unfavorable in that the problem is not partially solved or simplified, the response is suppressed and the next stimulus (perhaps a rereading of the problem) will elicit a different response. Whether or not this second response is correct, the learner is advancing toward his goal, for of all his repertory of possible responses he has eliminated one. Therefore, by simple trial and error, he progresses toward his goal. His learning curve will appear like this:

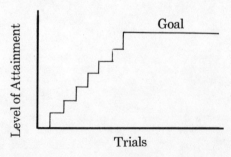

Figure 19. Student Responding by Trial and Error.

156

To be sure, the rate at which the goal is attained will vary from learner to learner. If past learning experiences have shaped his response pattern appropriately, his early responses may do much to reorganize and simplify the problem for an early solution. If not, his early responses may actually introduce spurious elements into the problem which may lead to a sequence of incorrect responses before they are resolved. The learning curve may, in general, follow one of the two basic forms shown below.

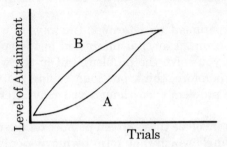

Figure 20. Students Influenced by Earlier Learning Experiences.

In case A, initial responses are incorrect and delay attainment while in case B correct initial responses lead to a relatively high attainment in a very short time.

But are these learning curves adequate to explain all problem-solving behaviors? Consider the following simple problem-solving experiment. Read the entire experiment before trying it with a subject.

Problem-Solving
Moves

Arrange fifteen pennies on a table so that they form a solid triangle as in the figure below. Read the following instructions to the subject.

"This is an experiment to determine the ways in which different people solve problems. I am not interested in the speed or cleverness with which you solve the problem, but in the way you arrive at a solution. Therefore, think out loud telling me whatever you think about the problem even though you may later change your mind."

"The problem involves these fifteen pennies on the table. You are to invert the triangle which they form by moving only five pennies. Do you understand the problem? All right, tell me what you would try to solve it."

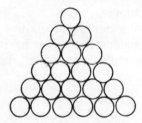

Figure 21. Penny Triangle for Problem-Solving Moves.

The solution of this problem is very simple once the subject realizes that at least some of the five moved pennies should be used to form the bottom vertex of the new triangle. But subjects who begin by removing pennies from the bottom row to form the new bottom vertex will likely make several attempts at the problem before achieving a solution. We want to record the pattern of attempts made by these subjects. To make such a record is relatively easy if you realize that there are only five correct positions to which pennies can be moved. The order in which these positions are achieved is not crucial to the solution of the problem. We can simply record the moves made by the subject as right (R) or wrong (W) by mentally matching them to the following pattern.

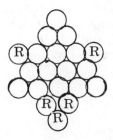

Figure 22. Penny Triangle Showing Correct Moves.

Pennies moved to any other position are wrong. Of course it may be that the subject will move a penny to a position (either right or wrong) and then change his mind. This type of move can be listed as C (for cancel). We are interested in the pattern of right, wrong, and canceled moves. This pattern should be recorded on a piece of paper held so the subject cannot see. A typical record might look like this:

W C W W C C R C W C R R W C R R

The problem should be solved when five right moves have been made without being canceled.

In practice a subject may make several consecutive moves followed by several consecutive cancellations. If these moves happen, our record should specify whether the cancellation involved a right move or a wrong move. This specification can be managed by using subscripts. C_R would indicate that a right move was canceled, C_W would indicate that a wrong move was canceled. With this convention, a problem solution might be recorded this way:

R W W R W $C_W C_R C_W$ R C_W W R C_W R W $C_R C_W$ R W C_W R

Using this scheme, record the problem-solving pattern for at least five subjects.

Analyzing the Data

We can use the problem-solving records to approximate the learning curves involved by using the following procedure. Divide a vertical axis into five equal intervals and a horizontal axis into as many intervals as moves that were made by the subject. Using the horizontal intervals, plot each move using a line of slope +1 for right moves, a line of slope −1 for the cancellation of right moves, and horizontal lines for wrong moves and the cancellation of wrong moves. Each segment should be joined to its preceding segment. Using this convention, the first example forms Figure 23a and the second forms Figure 23b.

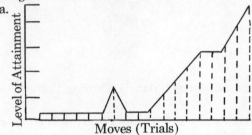

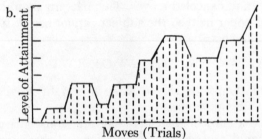

Figure 23a and b. Plotting Moves of Subjects.

Did you record any problem solutions which did not fit the learning curve patterns discussed in the first section? Perhaps you found several subjects who solved the problem on the first five moves. Did they do their analysis before they began to think out loud, or had they seen a similar problem before? What previous experiences had shaped their response pattern to yield a quick solution?

On the other hand, you may well have found a problem solver who showed no particular response pattern for many trials and then suddenly made five consecutive right moves. Learning curves like Figure 24 lead to a hypothesis of some kind of "insight" on the part of the problem solver. This speculation is a very attractive one which is difficult to measure and test, but one which many people feel captures the essence of their own problem-solving techniques. In the case of the pennies, this insight involves how the subject views the problem of constructing the "new" vertex (either within or outside of the original configuration). He may not be able to verbalize this insight, but at the same time, we should not restrict learning to the ability to verbalize.

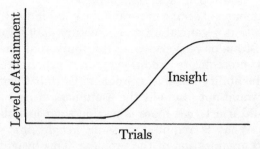

Figure 24. Subject With No Particular Response Pattern.

Gestalt Psychology

The insight that comes in solving the penny problem does not come by analyzing the configuration penny by penny, but by viewing parts of the pattern (the vertices) in relationship to the whole. This viewing certainly implies a much different orientation to learning than the atomistic approach followed by S-R theorists. It indicates that we should be concerned with the relationship of patterns and configurations to learning. This position is a branch of psychology known as *Gestalt Psychology*, originally proposed by the German Philosopher-Psychologist, Max Wertheimer. (The German word *gestalt* means organized pattern, configuration, or form.)

Wertheimer held that we see things as meaningful wholes, and we break them into small components only by very artificial processes. Thus, we see the dots printed below not as separate points but as whole figures.

Figure 25. Figure Perception.

The points tend to be seen as the critical points of the figures. The first group of points is identified as a triangle, square, pentagon, etc., while the last figures are more apt to be seen as circles. In fact, none of these figures are actually present, and the points of the "triangle" and "square" could lie on the circumference of a circle just as easily as the points of the "octagon."

The phenomenon of seeing meaningful wholes indicated to gestalt psychologists that organization and structure play important roles in learning. Stimulus-response theories imply that the possession of necessary prerequisite knowledge and experiences somehow guarantees the ability to solve a problem. Gestalt theorists, on the other hand, would not discredit the usefulness of past experiences, but they would insist that more is needed than just the necessary information. What is required is some insight into the overall structure and organization of the problem. This insight is possible only if all necessary aspects of the situation are available for observation. Master teachers are aware of differences between situations which facilitate understanding and those which seem to work against understanding, even though the same ultimate teaching steps are used and the same conclusion is reached. In the best situations, the teacher not only presents the steps but also arranges them so that the relationship between the most important features is emphasized and distracting information is minimized. At the other extreme are teachers who present all the correct steps but neglect to give an overview of where the steps are leading or what the articulating principles are. Their students often fail to gain insight into the problem or theorem because the basic structure or organization is effectively (though often inadvertently) hidden from them.

There are three types of evidence which careful observers often note in situations where insightful learning is believed to have occurred. The first of these is an interruption of movement for a period of time, a hesitation or pause for survey, inspection or concentrated attention which is often followed immediately by the solution of the critical part of the problem. Often the child will lay down his pencil, or even get up and walk quietly around the room, only to return to his desk and write down the required answer. Second, the solution can be readily repeated after the first critical solution. Once a subject solves the penny problem, there really is no "problem" anymore. He can invert the triangle in a minimum number of moves without repeating any of the trial and error attempts, or without the need for any practice. Third, a solution with insight can be generalized to new situations which have apparent relationships. Once the penny problem is solved the subject can find analogous solutions for other triangles, no matter how large or small the dimensions.

Gestalt psychologists hold that learning proceeds according to four basic laws of perceptual organization. These four laws are all related to an overall guiding principle: psychological organization tends to move toward a state of "good" gestalt. The properties of good gestalt are regularity, simplicity, and stability and are illustrated by the four laws.

1. *The law of similarity.* Similar items that are alike in form or color or in similar transitions (alike in the differences or steps which separate them) will tend to be grouped together in perception. Similar pairs tend to be learned more readily than dissimilar ones.

2. *The law of proximity.* Perceptual groups are formed according to the nearness of the parts. If several parallel lines are spaced unevenly across a page they will be seen as groups against an empty background according to their relative closeness to each other. Sounds will be grouped according to their closeness in time, and this organization or lack of organization will be reflected in learning, retention, and recall.

3. *The law of closure.* Closed areas more readily form figures in perception than unclosed ones. In learning, the direction of behavior will be towards an end situation which brings completeness or closure with it. A problem is perceived as being incomplete, and the achievement of closure is satisfying. In this manner solutions can serve as their own rewards.

4. *The law of good continuation.* Organization in perception tends to occur in such a manner that a line segment appears to continue as a straight line and a circular arc as a complete circle. The learning of paired figures is facilitated when one figure can be perceived as a continuation of the other. Both continuation and closure facilitate oganization, and organization, in turn facilitates learning.

The following article by George Hartmann illustrates these gestalt laws in terms of mathematics content and suggests their application to teaching.

Gestalt Psychology and Mathematical Insight

George W. Hartmann

This paper is presented as an expression of belief and hope that some of the newer formulations in psychological theory will bring about a democratization of mathematical competence somewhat akin to the astonishing elevations of performance often obtained in the field of reading. The symbolism and thought processes involved in using conventional language are not essentially different from those employed in the more universal terminology and operations of mathematics. If reading ability is now better and more widespread among Americans than at any previous time in our history, there is no reason why a better comprehension of numbers, figures, and conceptual relations should not make quantitative inferences as common as simple words and sentences. Although it is probably true that algebra and geometry *as ordinarily taught* are unable to become more influential because of the present intellectual and motivational limitations of the average learner, I should like to maintain the position that by taking more adequate advantage of the principles of mental development, our teachers could make the special modes of thought of the mathematician a part of the daily routine of our citizenry. The deserved success of the Bell and Hogben volumes reveals what some of these possibilities are.

Most persons will probably approve of this end and accept the claim just made, but many maintain that the existing crisis in secondary-school mathematics — a situation which affects more than the jobs of certain teachers of specialized subject-matter — is a consequence of altered social conditions and changed educational attitudes, and that a better methodology and modernized psychology can do little more than make possible a graceful retreat to a more humble role. The essential problem, these folks say, is a curricular one. True enough; but it would be a mistake to assume that there are no psychological foundations to a curriculum, mathematical or otherwise. The human values in both content and procedure are too intimately allied to permit that. It is a common error which holds that an educational discipline becomes "progressive," i.e., modern and enlightened, by skillfully and effectively teaching that which should not be taught and not teaching that which should. The *How* and the *Why* of instruction are organically related and a truly satisfactory solution for one will tend simultaneously to solve the other.

Reprinted from the October 1937 issue of *The Mathematics Teacher* (XXX, 265-70). © 1966 by the National Council of Teachers of Mathematics. Used by permission.

Insofar as any distinct psychological system has been adopted by mathematics teachers, their training appears to have led them to favor the connectionist or "bond" theory of learning, although the frank hostility of this position to the claims of the formal disciplinarians has led many of them to cling desperately to the long-discredited notion of separate mental "faculties." This occurred, one may suppose, not because mathematicians are any more conservative than other pedagogues, but because such an outlook supported their claims to a preferred status in the conventional course of study. In the last decade, however, a small but growing group has found a more satisfactory foundation for its practice in the tenets of the Gestalt brand of psychology — a theory to which mathematicians are perhaps temperamentally congenial because it literally outlines a subtle "geometry of the mind." What are some of the considerations upon which this advanced (and advancing) viewpoint rests?

In my judgment, there are three propositions which are basic to that type of theorizing which goes by the name of Gestalt:

1. *All experience or mental life implies a differentiation of the sensory or perceptual field to which the organism can respond into some kind of figure-ground pattern.* In other words, there must be heterogeneity of stimulation before any psychological process can occur. If we were affected by nothing but undifferentiated homogeneous energy, e.g., a single flat level of grey in vision or an unvarying tonal mass in hearing, the very conditions for consciousness itself would probably be absent. Difference produces phenomena. From this standpoint, variety is more than the spice of life — it is a prerequisite of life itself.

This dualism of figure and ground is an inescapable feature of all perception, but it is a *functional* and not a structural antithesis. The figure is simply that feature of the situation to which primary attention is given *at the moment* — the ground, although essential to the emergence of the figure, has a secondary role in terms of the focalization of

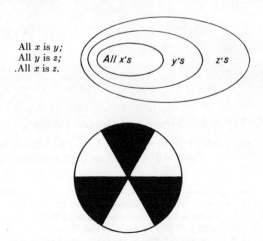

All x is $y;$
All y is $z;$
.All x is $z.$

All x's y's z's

Figure 26. Ground-Figure Relationship.

Note how the black and white propellers are alternately seen.

the organism's interest. In Figure 26 (which is typical of all "reversible" patterns), the black and the white areas alternate in dominating the reader's field; when the black region is figure, the white is ground, and vice versa. Most of the patterns we encounter are far more stable than this, although all can be reorganized subjectively and made to fluctuate to some extent. Thus, the interlinear white space on this page, which is normally "unnoticed" ground even though it is absolutely essential to the reading act, can — with some effort — undergo a transformation and acquire temporarily the status of "figure."

2. *The course of mental development is from a broad, vague, and indefinite total to the particular and precise detail.* The end-result of this process of differentiation is an organized body of "clear and distinct ideas" — that state of affairs so dear to the mind of the skilled logician. But it is far from the condition with which the growth process starts. Sharpness of outline is what we end with, not what we have at the beginning. The act of perceiving is normally initiated by a dim general awareness of the object; it is only as this continues to act upon the observer that its internal "structure" emerges.

In the light of this conception many mathematical commonplaces acquire a new meaning. Euclidean geometry, e.g., is a masterpiece of reasoning, but it would be far truer to the facts of genetic psychology if the order in which it is commonly presented were completely reversed. Its logic is atomistic or elementaristic,[1] i.e., it begins with the most

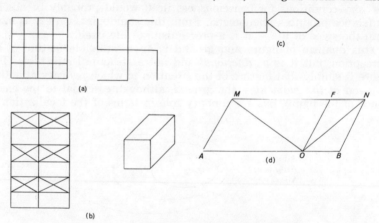

Figure 27. Figure and Outline Perception.

How "field forces" govern what is discriminated. Can you isolate the swastika and the "box" in their companion figures on the left?

---

[1] The following clipping from the humorous column of a teachers' journal indicates that many persons have become aware of the artificial and "absurd atomism" implied in many of the textbook problems of an older day.

"Arithmetic is a science of truth," said the professor earnestly. "Figures can't lie. For instance, if one man can build a house in 12 days, 12 men can build it in one."

"Yes," interrupted a quick-brained student, "Then 288 will build it in one hour, 17,280 in one minute, and 1,036,800 in one second. And I don't believe they could lay one brick in that time."

highly refined and mature abstractions, such as "definitions" (note the term with its suggestion of optical focusing!) of *point* and line, rather than starting with more massive and "natural" percepts like cubes, surfaces, etc. Psychological experimentation indicates that a "point" is a fairly late and high-grade achievement of one's spatial understanding. Our visual-tactile world is not originally made up of *points* — instead, these emerge from it. Paradoxical as it may seem in the light of the usual placement of courses, solid or tri-dimensional geometry is the source of all later spatial analysis which ends in, but does not proceed from, the strange entity that "has" location without extension. It has even been argued that division is a more primitive arithmetical operation than addition.

3. *The properties of parts are functions of the whole or total system in which they are imbedded.* In perception this principle is clearly observed by the fact that a grey square upon a blue ground looks yellowish and the *same* grey patch on a yellow field appears bluish (color induction or "contrast"). In Figure 27(c), most persons "see" a square and a hexagon in contact (presumably because the structural organization of the drawing favors this response), but the capital letter K, which is just as much present in substance, is usually not discerned. To isolate a familiar portion of the alphabet in this situation requires the segregation of two markedly dependent parts of two strongly unified "figures." Frequency and repetition cannot account for this phenomenon, for even the most experienced geometer has encountered a K more often than he has seen these elementary patterns in contact. The external arrangement as given compels a corresponding internal organization.

Figure 27(d) is even more impressive because it exhibits some of the mechanism underlying an illusion. Most persons, if asked to compare the diagonals MO and ON, will unhesitatingly declare MO the longer. Actually the two are drawn of equal length. The effect is apparently traceable to the larger rectangle $MAOF$ which causes the observer to "see" (not to "infer" in the usual logical sense) its diagonal as greater than that of the noticeably smaller rectangle $FOBN$. The "illusion" is partly overcome by erecting a perpendicular at O, thus minimizing the unanalyzed and unequal influence of the two major areas. A better example of the Gestaltist's claim that a line is functionally a derivative of a plane could hardly be found.

It seems probable that the meaning of a number in series is a special case of membership-character being conditioned by its role in some structure. Thus, the number "364" is comparatively meaningless in isolation. Conceptually, however, its fuller meaning is necessarily derived from some schema, such as "less than 400," "between 350 and 400," "nearer 400 than 300," etc. In the case of lightning calculators, most numbers have acquired some such "individuality" as this — a fact which contributes something to an understanding of their ability.

With these three generalizations as one's conceptual tools, it is surprising how many obscure phenomena swiftly become more intelligible. A number of years ago, while examining the arithmetical errors of college students, I noticed that a mistake was more likely to occur when a larger number was being added to a smaller one than in the converse case. Thus, $9+7$ and $7+9$ both make 16 but addition errors are decidedly more frequent with the latter combination. The rule has also been statistically verified for two-place numbers, and I am inclined to believe it holds for fractions and any other number combinations. If we view the addition of two quantities as a simple case of completing an indicated total, then this observation is brought under the head of "closure" phenomena. In any language completion test, gaps are to be filled in, and the test's difficulty is roughly proportional to the number and extent of the gaps involved. This "totalizing effect" is seen in the figures below. A "circle" with three-fourths of circumference visible is easily seen as a full circle at a slight distance from the eye — a fact occasionally used by the oculist in visual testing. An arc equivalent to a quarter-circle does not lend itself so readily to the "restoration" of the entire circle (cf. Figure 28). *Pari passu*, when one adds 9 to 7, one is traversing a greater psychic distance than when one adds 7 to 9. Hence, the goal, "16," is more surely reached in the latter instance. Since there is no inherent difference in the difficulty of the two symbolic number combinations, the variation in accuracy must be traced to the

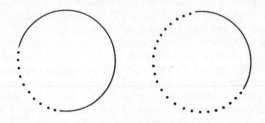

Figure 28. Closure Effects.

Closure effects are easier, less ambiguous, and more impressive the less the perceptual "filling" required.

forces behind the schematic structure by which they are represented in the organism.

If this explanation makes concepts obey the same laws as percepts, that has its source in the conviction that thinking and reasoning are dependent upon the processes of perceiving. The latter activity is nearer to the concrete situation than the former, and ordinarily involves less of the inferential closure represented by the dotted portions of Figure 28. But in the case of "Euler's circles" as used in elementary demonstrations of formal logic, one literally "sees" how intimately syllogistic proof is linked to direct sensory perception of the basic pattern. It seems that the famous Swiss mathematician of the eighteenth century was once a tutor by correspondence to a dull-witted Russian princess and devised this method of convincing her of the reality and necessity of certain relations established deductively. Thus, the syllogism can be tested for its truth or falsity if diagrammed as indicated. Here the concentric "ellipses" reproduce the essentials of the situation so faithfully that the answer becomes a matter of "mere inspection."

This process of making an organism aware of the conditions governing the phenomena to which it is reacting is essentially what is meant by the "insight" experience. Rightly construed, insight is not a peculiarity of the "higher" rational functions, but a process necessarily occurring at all mental levels. Simple adjustments of bodily position, such as lifting the foot to go up a step or bending the knees when accepting an invitation to sit down, constantly exhibit a low but decidedly real level of insight. In every case there is a rearrangement of an action pattern as a result of changing forces in the "field." If the new organization fits the needs of the situation, the neural stresses and strains return to a state of relative dynamic equilibrium and the "problem" is said to be "solved."

A simple geometrical example will suffice to clarify this all-important point. Given Figure 29(a), a high school sophomore is asked to find the area of a circumscribed square, knowing only the radius of the inscribed circle. A question or assignment such as this normally produces

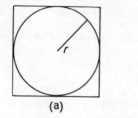 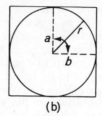

(a)   (b)

Figure 29. Quantitative Relations.

How insight operates with quantitative relations.

a mild tension which does not vanish until an answer satisfactory to the organism is achieved. If one examines carefully what occurs, it is plain that a state of mild bewilderment sets in which lasts until a configuration emerges. Figure 29(b) is typical of a wide range of educational situations. The pupil is blocked as long as he sees the key item (the radius r) *fixed* in its first position; but as soon as he *shifts* it mentally to position a or b, its properties are *transformed* — it is no longer a half-diameter, but is now half of a side of the square. This reorganization accomplished, the response $A = (2r)^2$ is immediately and confidently made. The delay involved is almost entirely taken up by the time required to bring Gestalt principle No. 3 (above) into action.[2]

Another basic conception of Gestalt theory which promises to be useful in mathematical instruction is the idea of transposition. This had its origin in the common observation that a musical melody is not the same as the sum of the separate tones that presumably comprise it. "The Star Spangled Banner" can be sung by bass, soprano, and other voices, or played by various orchestral instruments in different "keys." In none of these instances need any of the individual tones be alike, and yet the common pattern or melodic sequence is easily recognized. The melody is the whole which is transposed and *transposable* from one situation to another.

---

[2]Alternative and equally simple solutions are also possible.—On one occasion a mathematics teacher, who was listening to an *oral* exposition of this example, insisted the answer was wrong because he had heard and interpreted $(2r)^2$ as $2(r)^2$. It is significant that erroneous as well as correct solutions are equally intelligible under this principle.

Is it not conceivable that most of the "abstractions" with which mathematics deals are of this nature? The ratio between the diameter of any circle and its circumference is expressed by the constant π, and this relation is transposable whether the actual concrete materials that constitute the circle are made of copper, rope or carbon and liquid particles. In the case of the equation — which is plainly the heart of most mathematical operations — we see how the procedure is built around the possibility of preserving "permanence amid change." One may think of the equation as a step-wise expanding or contracting pattern that preserves invariant the condition of equality or identity throughout the various transformations that may be legitimately applied to it.

This situation was long ago appreciated by gifted thinkers and accounted for much of the esteem in which mathematical method was held. However, it would be a great mistake to assume that other aspects of nature and behavior are lacking in this respect.[3] Even a goldfish (who is known to be less intelligent than the humble cockroach) reacts appropriately to a group of "equivalent stimuli." Thus, suppose he has learned that food is to be found behind that light which is intermediate in intensity among three sources of illumination in his field. Now change the absolute brightness of the three bulbs by doubling or halving the "intensity" of the current. In either event, the goldfish swims unhesitatingly toward that light which is relatively in the "middle" of the triad. This response — found widely in all species — is difficult or impossible to explain on the basis that learning is specific, i.e., restricted in its effectiveness to the precise stimulus-object employed. This interpretation itself must be false as it is killed by its own hyper-specificity. No two learning situations can be alike in *all* respects; the likeness is found in the organization of the wholes and not in the substance of the pieces. A wooden table is functionally more like a metal table than it is like a wooden chair. The way things are put together determines the attributes of the system thus established.

It must be obvious from these illustrations that mathematical and psychological research have more in common than is usually realized. The field and organismic approaches to behavior have more than a purely physical and biological connotation, and point set theory is heavily used in systems of "topological" psychology. Much of this is the inevitable and desirable consequence of the unity of scientific thought. The contributions of mathematical technique to psychological advance have been most impressive, but the reverse type of obligation has rarely been incurred, presumably because there was so little to borrow! Perhaps if mathematics teachers act upon the recognition that the content of their discipline has to be rediscovered and created *de novo* by every learner, they will have provided themselves with the one tool needed to make the Grand Tradition of rigorous thinking influential in the lives of our people.

---

[3]Consequently geometry cannot be the only area in which rigorous proof is possible. No proof can rise above the assumptions and hidden theorems upon which it rests.

Heuristic Teaching
in Mathematics

Mathematics as Activity

Hartmann concludes his article with a remarkable statement. "Perhaps if mathematics teachers act upon the recognition that the content of their discipline has to be rediscovered and created *de novo* by every learner, they will have provided themselves with the one tool needed to make the grand tradition of rigorous thinking influential in the lives of our people." (5:661) What a radical idea! The implication here is that the learner should *make* mathematics, not just receive a full-blown creation. Is mathematics something that children can make, flexible, dynamic and accessible? Or is mathematics a static imposing edifice which we receive as an inheritance? Certainly mathematics is more than patterns to be memorized, concepts to be acquired, or facts to be learned. Mathematics is something to *do*. More than a static accomplishment, mathematics is a process, an activity, a problem-solving technique. These problems may be mathematical models of physical situations, or they may be problems of organization and pattern (problems of "pure mathematics"). But no matter what the exact nature of the problem, most people would agree that what we ultimately try to teach in mathematics is the art of problem-solving.

This aspect of mathematics is exactly what should be reflected in mathematics teaching. Certainly the work of Gestalt psychologists provides insight as to how problem solving proceeds for many individuals. But it is somewhat less helpful in suggesting strategies for attacking the problem. Such strategies are called heuristics, and teaching which emphasizes these problem-solving approaches is known as heuristic teaching. Many people equate heuristic teaching and discovery teaching. But to a mathematician the word, *heuristic*, has a much richer meaning than simply discovery. The foremost advocate of heuristic today, George Polya, reminds us that:

Heuristic, or heuretic, or *ars inveniendi* was the name of a certain branch of study, not very clearly circumscribed, belonging to logic, or to philosophy, or to psychology, often outlined, seldom presented in detail, and as good as forgotten today. The aim of heuristic is to study the methods and rules of discovery and invention.

Modern heuristic endeavors to understand the process of solving problems, especially the *mental operations typically useful* in this process. It has various sources of information none of which should be neglected. A serious study of heuristic should take into account both the logical and psychological background; it should not neglect what such older writers as Pappus, Descartes, Leibnitz, and Balzano have to say about the subject, but it should least neglect unbiased experience. Experience in solving problems and experience in watching other people solving problems must be the basis on which heuristic is built. (8:112, 129-130)

What Polya seems to be implying is that heuristic includes discovery — and much more. In mathematics especially, heuristic seems inexorably bound to problem solving. Polya has devoted the major portion of his writing and lecturing to an explanation and analysis of problem-solving techniques. Much of his writing can be characterized as case histories of solutions. The most concise formulation of the heuristic which he has synthesized from these many case histories is contained in his most popular book *How To Solve It*. He presents the heuristic as a list of questions which one should ask himself as he tries to solve a problem. What is important in Polya's heuristic is not simply asking questions, but the logical strategy behind the *kinds* of questions he asks.

This point has often been overlooked. The old concept of heuristic teaching tended to focus on teacher-asked questions but to ignore the kinds of questions that were asked or the logic behind them. We need to take another look at heuristic teaching and to relate it as closely as possible to the meaning of heuristic in mathematics. One possible way to arrive at this viewpoint is to define heuristic teaching in mathematics as *a category of instructional methods which make primary use of one or more problem-solving techniques in mathematics.*

Now techniques of problem solving are only minimally relevant apart from problems. If our definition is to have any import at all, we must assume and find support for the assumption that a nontrivial part of school mathematics content can be approached as if it were a problem. This requirement alone may require a major shift in the teacher's philosophy and view of the teaching situation. If content is to be approached as if it were a problem, then the classroom must change from a place where information is provided to a place where information is sought. A student has a problem when he has the description of something but does not yet have anything that satisfies the description: 1) he has a clearly defined goal that he desires to attain; 2) the path toward the goal is blocked and his habitual responses and fixed patterns of behavior are not sufficient for removing the block; and 3) he can discriminate between alternative courses of action and deliberate about their feasibility. (6) To approach mathematics content as a problem, the mathematics content must be established as a goal related to possible courses of action. The forming of the goal is related to the establishment of objectives, the related courses of action to the determination of hierarchical prerequisite knowledge.

Content-as-Problem

How can we approach mathematics content as if it were a problem? The *form* of the statement of objectives seems particularly important to the establishment of content-as-problem. A statement like "today we will learn how to compute the distance between two points in a Cartesian coordinate plane by using the Pythagorean theorem" destroys any content-as-problem approach. It establishes a goal but then proceeds to remove the blocking of that goal by removing the need to discriminate between various courses of action. In this case, a far simpler goal statement seems much more preferable: "let's see if we can figure out how to compute the distance between any two points in a Cartesian plane."

Much of the ability to present content as problems depends upon just this ability to break mathematics into "let's-see-if-we-can-figure-out" blocks. This breakdown is most easily done by asking the student to redevelop or recreate many of the principles and relationships which form the body of mathematics. This recreation was a formidable, if not impossible, task when the teacher used traditional mathematics curricula. These curricula viewed mathematics as little more than a collection of facts and principles. But modern mathematics curriculum projects have worked to show the logical connections between mathematical definitions, concepts, and principles. As a result, the possibility of presenting mathematics content as problems has come within closer reach.

But the concern of heuristic is *not* just obtaining an answer to a problem. Heuristic seeks generalities from problem solutions. The last category of Polya's *How To Solve It* list is entitled "looking back." He admonishes the problem solver to "*examine* the solution obtained. Can you check the result? Can you check the argument? Can you derive the result differently? Can you see it at a glance? Can you use the result, or the method, for some other problem?" (8:xvii)

This looking back is often ignored in the usual problem-solving sessions in many mathematics classrooms. The result is an almost totally answer-oriented student. He does not *want* to know about the structure of mathematics. He cannot be bothered with basic laws and principles. He will only grudgingly tolerate explanations of why a particular mathematical procedure works. We have conditioned him to a belief that, in mathematics, answers are all-important. Answers are almost always the goal of mathematics tests, not discussions nor explanations. He seeks not the "why" of a solution but the "how," the tricks, the manipulations.

Presenting content as problems, then, is not the same as presenting a series of answer-seeking exercises. The content-problems of heuristic teaching must look back over content, techniques, and concerns that have previously entered into a student's experience. They must look forward, as well, to new relationships and new problems. What does this hindsight and foresight mean in terms of lesson design and sequencing for heuristic teaching?

Logical Construction and Instructional Procedures

Heuristic teaching has been considered by some mathematics educators as a teaching procedure which makes use of a series of directed questions. "By (heuristic teaching) we mean a method which aims to lead the student, through well-chosen questions, to discover facts, information, relationships, and principles for himself." (1:167) This confusion may well have arisen from the question of whether Polya's lists of questions formed a method or a goal. (Indeed, much of what makes Polya a master teacher is this intimate intermingling of method and goal.) The difficulty is compounded by the fact that no criteria or guidelines were established to select or evaluate "well-chosen questions" or well-chosen sequences. As a result, the old concept of heuristic teaching was only minimally related to problems or to problem-solving techniques. The following excerpt from the 21st Yearbook of the National Council of Teachers of Mathematics is an example of what was considered heuristic teaching.

First lesson on circles: If you wished to construct another circle equal to this one, what would you measure? Consider this circle with the 5 inch radius. With respect to the circumference, where would a point 3 inches from the center be? 8 inches? 5 inches? What are your conclusions with respect to distance from the center and the circumference? Here are two equal circles, 0 and 0'. Mark off equal arcs AB and A'B' and draw the radii. What would you expect to be the relationship be-

tween angles AOB and A'O'B'? What is one method of proving two
angles equal in two equal circles? What do you think is true about
chords AB and A'B' in these circles? How do you usually prove two
line segments equal? But there are no triangles here! How would you
draw lines to make the triangles which you mentioned? (2:317)

Consider the sequence of questions in this example of heuristic
teaching. Do they bear any relationship to the list of questions that
comprise Polya's heuristic? To be sure, we are faced with a problem: how can a given circle be duplicated? Polya would begin by
exploring the given data and conditions; he would continue by
looking for similar problems (have you solved problems where other
geometric figures were duplicated?); and he would try a simpler
problem (how can you construct a circle if the exact size is not important?). The questions in our example bear little relationship to
any general problem-solving techniques. Teaching by simply asking
a series of direct questions would, of and by itself, certainly not
seem worthy of the designation "heuristic teaching." *What we
should seek in heuristic teaching is a relating of the logic of the
teaching sequence to the logical patterns of problem solving.* Consider these patterns to see how they can determine logical instructional sequences.

I. One problem-solving technique involves guessing an answer,
working out its consequences, and by comparing these with the
original conditions of the problem, improving the original guess.
The implications of this heuristic for classroom teaching strategy
are fairly obvious. Suppose we want to teach students how to square
a binomial. We want an equivalent name for $(x+3)^2$. We ask for
a guess; a common response would be x^2+9. But now we must look
for the consequences of our guess. Suppose the variable x has the
value 1. Then the value of

$$(x + 3)^2 = (1 + 3)^2 = 4^2 = 16$$

and of

$$x^2 + 9 = 1^2 + 9 = 1 + 9 = 10$$

Here is a case where these two expressions do not name the same
number. x^2+9 is somehow "too small." Can we improve our guess?
How about $(x+3)^2 = 2x^2+9$? Then, if $x=1$

$$(x + 3)^2 = (1 + 3)^2 = 4^2 = 16$$

and

$$2x^2 + 9 = 2 \cdot 1^2 + 9 = 2 + 9 = 11$$

This solution seems a *little* better. How about $(x+3)^2 = 7x^2+9$? Then, if $x = 1$

$$(x + 3)^2 = (1 + 3)^2 = 4^2 = 16$$

and

$$7x^2 + 9 = 7 \cdot 1^2 + 9 = 7 + 9 = 16$$

Success! But what if we vary the problem slightly? Have we gained any insight which would let us expand expressions like $(x+4)^2$ or $(x+178)^2$? And what happens if the variable x has a value other than one? These variations will lead to many more modifications of our original guess before the problem is solved.

II. Sometimes a problem-solving technique involves finding a simpler related problem. In some cases this simpler problem may actually be a part of the original problem with certain conditions ignored. This heuristic adapts itself to instructional strategy particularly well when the general objective is the exploration of a relatively wide area of topics. Suppose we want to teach a sequence of theorems about secants, tangents, and chords of circles. Many of these ideas can be subsumed in the problem "given three non-collinear points, can we find a way to construct a circle which will pass through these three points?"

To the uninitiated, this problem is not a trivial one. A trial-and-error approach will usually require many attempts before solution, and if a circle is fit to the points, the process usually does not suggest a definite procedure. What we need is some way to locate the center of the required circle. But suppose we take a simpler related problem. Can we construct a circle through any two points? Com-

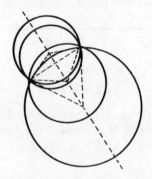

Figure 30. Problem-Solving Technique.

The 3 point problem can be solved as two 2 point problems by considering the pattern of possible circles.

pare the solutions of all the students in the class — are they identical? Do they form a pattern? Suppose we put two or three of these simple problems together. After all, in our original problem of three points, we have three possible combinations of pairs of points. What arcs and lines do the patterns suggest; what relationships seem to exist between these arcs and lines? Can we establish these relationships deductively from what we already know?

III. Polya discusses decomposing and recombining as important mental operations and as important techniques in problem solving. We can often consider a problem as a complex situation made up of many details. We begin by focusing upon the details, individually decomposing the whole into its parts. In this process, it may be necessary to go back to the original definition of a term and to introduce new elements involved in this definition. We then attempt to reorder and recombine our original and new elements in some new and different way.

This problem-solving technique is useful in designing instructional procedures in which the ultimate aim is one of classification or definition on the basis of a classification. Suppose, for example, that we wish to explore the definition of similarity in geometry. We could generate a set of triangles by considering many different images of a given cardboard triangle projected by an overhead projector at varying distances and angles from the screen or blackboard. The problem to be posed is to form sets of triangles and then to describe some criteria for including or excluding triangles from this set. The ensuing class discussion should proceed along the decomposition and recombining approach in a very natural manner, eventually generating the essence of the similarity definition.

Students will sometimes stop short of an accepted definition when this technique is employed. When this happens, the teacher should not force the accepted definition upon students but lead them to explore the consequences of their own defintion. Studying the consequences of a particular answer is always a good problem-

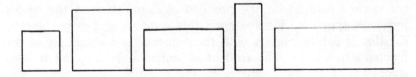

Figure 31. Figure Definition.

Are these rectangles similar?

solving technique. For example, students may be satisfied with a statement like "two figures are similar when corresponding angles are congruent." This statement is not incorrect, but it defines quite a different kind of "similarity" than that usually meant by mathematicians. By this definition all of the rectangles in Figure 31 would be "similar." (What would happen for five-, six-, or seven-sided polygons?) The choice here is not a matter of deciding between a right or wrong answer, but deciding whether or not to accept the consequences of that answer.

These examples affect the organization of instructional sequences. They provide the beginning of a set of logical principles for the construction of classroom lessons and instruction. The logic of teaching but the actions of teaching as well. What do our exam-come to fruition, such a theory must encompass not only the logic of teaching, but the actions of teaching as well. What do our examples say about the actions of the teacher during heuristic teaching? At first inspection they make little or no determination of action. One can deliver a lecture in which the problem-organized content is developed by reference to simpler related problems just as well as a student-centered discussion. A similar statement could be made about each of the other examples.

But heuristic does imply some determination of instructional action as well as logic. Therefore, the basic meaning and principles of heuristic must be reexamined. The following section compares the original example of heuristic teaching from the NCTM 21st yearbook with a statement about heuristic from the field of computer simulation of human problem solving.

Uncertainty and Heuristic Teaching

Look again at the first lesson on circles (page 168) which served as an example of the old notion of heuristic teaching. It has little to do with problem-solving techniques, as we have seen. But it differs from the instructional strategies we have just discussed in yet another important way. That the teacher has predetermined the correct solution to the problem of duplication in the circles lesson is apparent. Furthermore, this solution and only this one solution is determined by both the sequence and specificity of the teacher's questions. The student is immediately directed to consideration of metric geometry in the question "what would you measure?" The radius is specified by the teacher in the example used.

To simply consider heuristic teaching as a fancy phrase for discovery teaching is not sufficient. Most of discovery teaching could (like our example) be more correctly described as "uncovery teaching." An inductive sequence of steps is constructed around a particular problem solution or concept organization. If the teacher knows of only *one* means to the end, the task of the student becomes one of uncovering this particular solution or organization. The option of unusual solutions or organizations is not entertained, and the discovery lesson degenerates into a game of "guess what's on my mind."

But the freedom to consider alternate possibilities lies at the heart of heuristic. This freedom has been recognized as a cardinal principle even in the area of research in the computer simulation of human problem solving. Geleunter and Rochester put it cogently:

"We shall consider that a *heuristic method* (or a heuristic, to use the noun form) is a procedure that may lead us by a shortcut to the goal we seek or it may lead us down a blind alley. It is impossible to predict the end result until the heuristic has been applied and the results checked by formal processes of reasoning. If a method does not have the characteristic that it may lead us astray, we would not call it a heuristic, but rather an algorithm." (4:337)

How different this explanation is from the usual teaching procedure where an avowed purpose of the teacher is to *prevent* children from going astray! *If we are to make heuristic teaching compatible with heuristic, our teaching must not only allow, but demand, the flexibility to entertain uncertainty and alternate solution approaches.* In practice, this area of heuristic teaching may be hardest to achieve.

The "unusual solutions" option necessary for heuristic teaching appears to be particularly hard to incorporate into written text materials. Textbooks which utilize a discovery approach tend, paradoxically, to lock both the student and the teacher to a programmed sequence of steps. The provision of alternate discovery sequences is an overwhelming task. (One mathematics curriculum group estimated that to provide a reasonable number of options for a first-year algebra course would require a text of 50,000 pages!) Enough steps must be provided to insure that the student has the necessary prerequisite knowledge so important in mathematics. Yet, possible shortcuts cannot be clearly marked for fear of giving more authoritarian guidance.

When a teacher can face uncertainty, the results can be astounding. In an article in *The Arithemtic Teacher*, Joan R. Needleman describes a seventh grade lesson on locating points in a plane. Given a point on the blackboard, and faced with the problem of describing its location, a student made the unexpected suggestion that it could be located between regions determined by two oblique coordinate lines as suggested in Figure 32.

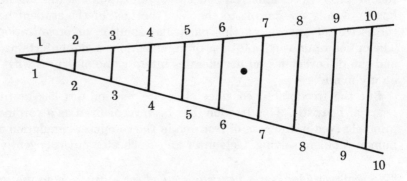

Figure 32. Point Location.

The point is located in the region determined by the 6th and 7th units.

Obviously, a teacher who is concerned with preventing children from missing the correct solution would have stopped the procedure at this point. But on the basis that "mathematics is a game, and that mathematicians can make different sets of rules and then play the game according to the rules selected," this teacher let the class proceed. The next student suggestion was the following figure, locating the point on a particular ray.

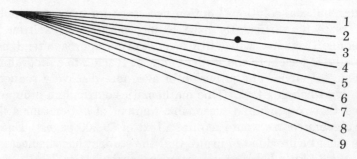

Figure 33. Point Location.

The point is located on ray 3.

The final suggestion was that two kinds of information were needed to locate the point: the number of the ray, and the distance of the point from the origin of the ray. (7)

By accommodating uncertainty this teacher was rewarded with a very successful, but most unusual solution! This freedom of response is a most necessary condition for heuristic teaching.

Heuristic Teaching and Student Involvement

One way to increase the probability of alternate solution approaches is to increase the input of ideas fed into the problem situation. This method argues, in turn, for involving as many different sources of ideas as possible. Some authors distinguish between teaching methods which consider the class as a whole and methods which point to students as individuals. (3:33) Only the latter are termed "heuristic"; group processes are called "genetic." There is very little in the nature of heuristic that would indicate that this distinction is a useful one. To be sure, problem solving in mathematics may be an intensely personal matter; yet, at the same time, few problem solvers would deliberately remove themselves from the input of other great thinkers and writers. Few mathematicians would be willing to forego opportunities for contacts with either their libraries or their colleagues! Why should the heuristic teacher deny students the opportunity to consult with their libraries or their colleagues? The kinds of ideas offered by a fellow student will be rough and unsophisticated, but can we be sure that the student is any better able to use a precise, powerful, and sophisticated idea from a mathematician? The sophisticated idea of the mathematician is treasured for its brilliant logical leap; a much more pedestrian and detailed development of ideas may well be more suitable for pedagogical purposes. The searching process common to so much of heuristic would argue that heuristic teaching should make great use of group processes.

At the same time, problem solving is, in an ultimate sense, an extremely individual matter. The mathematician, R. L. Moore, points out that no one of personal integrity who is in the process of solving a problem wants to be given the solution to the problem outright as done by another person. Moore invites and encourages his students to leave the classroom if they are not ready to see the solution of the problem under discussion. His students have been known to run frantically from the classroom, hands on ears, and not to return for weeks until they could reappear with their own solution to the problem in hand. If we are to pursue the matter of heuristic teaching, we must be open to such unorthodox activities. We must especially be prepared to abandon the rather preposterous assumption that the most valuable problem-solving activities can only occur within the confines of a classroom. And we must realize that problem solving is ultimately an individual process. The important question is how we can get students, as individuals, involved in the problem at hand. *Heuristic teaching seeks to maximize student action and participation in the teaching-learning process.*

There is *some* probability that a student will become involved in almost any kind of teaching act. However, to assume that the value of this probability rises in direct proportion to a rise in the extent to which students overtly participate seems reasonable. Thus, while it is entirely possible that a student will become caught up in a skillfully delivered lecture, the chances of his involvement would seem to be much greater (though not certain) if he is manipulating physical materials in a mathematics laboratory. Similiarly, working together to solve a problem with a small group may, for some students, result in more involvement than being isolated to struggle with the problem by themselves.

Summary

What is heuristic teaching? It has four important characteristics. Heuristic teaching:

1. approaches content through problems.
2. reflects problem-solving techniques in the logical construction of instructional procedures.
3. demands the flexibility for uncertainty and alternate approaches.
4. seeks to maximize student action and participation in the teaching-learning process.

Any teaching technique which meets *all* of these criteria may be called heuristic teaching. Certainly, heuristic teaching is not easy, and, in fact, there are many times when heuristic teaching is not even possible in mathematics. In particular, characteristics 1 and 3 seem to be especially restrictive. That all mathematics content can or should be approached as a problem is far from clear. How, for example, can the definition of the trigonometric functions be approached as a simple problem? Nor does it always seem possible to allow for alternate solutions in lesson planning.

Do these restrictions mean that we should abandon heuristic teaching? Heuristic teaching, like mathematics laboratories or discovery teaching, is a single technique for a teacher's repertoire. Just as different learning theories are necessary to account for the vast complexities in human learning, so different teaching techniques are necessary to deal with both human differences and different topics in mathematics. The search for one correct teaching style, like the search for one correct learning theory, is a search that is doomed to failure.

Nevertheless, heuristic teaching is an important teaching style and one that every teacher should try to master. The justification for the importance of heuristic teaching has been stated most eloquently by Polya:

A great discovery solves a great problem but there is a grain of discovery in the solution of any problem. Your problem may be modest; but if it challenges your curiosity and brings into play your inventive faculties, and if you solve it by your own means, you may experience the tension and enjoy the triumph of discovery. Such experiences at a susceptible age may create a taste for mental work and leave their imprint on mind and character for a lifetime. (8:v)

4-4: Study Module

For Further Investigation and Discussion

1. Some subjects learn the unusual nonsense syllables in investigation 3-1 before they learn the meaningful words. Can you use the Gestalt notions of figure and ground to explain this result?

2. Hartmann believes that the course of mental development is from a broad total to particular detail, and he makes some applications to geometry. How do his views compare to those of Piaget in regard to the "natural" order of geometry for young children?

3. Hartmann observed (circa 1937) that fewer errors were made when college freshmen add seven to nine than when they add nine to seven. For more recent research using first graders, see the article by Suppes, Patrick and Guy Groen. "Some Counting Models for First Grade Performance Data on Simple Addition Facts." Chap. 4. *Research in Mathematics Education.* Washington, D. C.: National Council of Teachers of Mathematics, 1967. In what ways do these two observations agree? What differences would be attributable to age difference between college freshmen and first graders?

4. Read the article by Berlyne, D. E. "Recent Developments in Piaget's Work." *British Journal of Educational Psychology* 27 (1957): 1-12. In this article Berlyne relates the position of Piaget to that of the Gestalt psychologists. He distinguishes between perceptual processes (studied by the Gestaltist) and thought processes and relates the stages in Piaget's theory to the relative degree that the developing child incorporates each of these into his intellectual functioning. What does this distinction imply for teaching problem solving in mathematics to young children?

5. The sharp rise in learning curves that accompanies Gestalt insight can also be explained by considering learning as an all-or-nothing phenomenon. Such an explanation is contained in Suppes, Patrick. "Mathematical Concept Formation in Children." *American Psychologist* 21 (1966): 139-150. Read the article. Can you explain your results for investigation 4-1 in a similar manner?

For Lesson Planning

Select a chapter in a mathematics textbook. Identify the basic idea of each section of the chapter. How many of these could be introduced by considering an appropriate problem?

Write a sequence of problems that could be used to teach those sections that lend themselves to heuristic teaching. Identify the major problem-solving strategies to be used with each section. Then list what a student would need to know before he could solve each problem. How does this list compare to a prerequisite hierarchy?

Which sections of the chapter you chose could not be introduced by considering an appropriate problem? Does the textbook you chose lend itself to heuristic teaching?

For Microteaching

An important skill for any mathematics teacher is the clear use of examples. Examples are useful in heuristic teaching, in basic explanation, and in general inquiry. Even when we ask students to put a homework problem on the board, we are asking for an example. Microteaching is a good way to practice giving mathematical examples.

The key to the good use of examples is organization. Organization is especially needed when the example is written out on the blackboard. Using the board requires thinking ahead. Is there enough space for all the problem? If not, be sure to organize the work in a series of adjacent columns. Few things are more confusing than an example that jumps all over the blackboard. Do you need a diagram to show relationships? Be sure to label it clearly. Does your solution involve several steps? Leave space between steps. This will allow you to go back and fill in fine detail as the class may request it. You may also want to leave space just to the right of each step for procedural notes, explaining what action was taken to proceed from one step to another. Remember that the reasoning between steps is usually more important than the steps themselves. Be sure to label the answer clearly. Relate it to the diagram, or the original problem statement.

You may want to use an overhead projector instead of a blackboard. The biggest advantage of an overhead projector is that you can face the class while writing the example on the top of the projector, making it easier for you to respond to student difficulties or questions when they arise. The overhead projector also makes possible preparation of written work ahead of time on a sheet of plastic. When students need to have lots of homework problems explained, you can save class time by assigning one problem to each student to prepare on a sheet of plastic ahead of time.

Prepare a problem to be presented as an example in a microteach. You may use either the blackboard or an overhead projector (if time permits you might want to use both). Have your presentation recorded on videotape if at all possible. When the presentation is over, view the tape and examine your organization. Are your steps clear? Is your strategy for solution obvious? Did you become so wrapped up in your own explanation that you ignored student questions? If necessary, improve your example and present it again.

For Related Research

The Gestalt psychologist is interested in problem solving, and problem solving is a major research area in mathematics education. An excellent review of research in this area is found in Kilpatrick, Jeremy. "Problem-Solving and Creative Behavior in Mathematics." *Studies in Mathematics, vol. XIX: Reviews of Recent Research in Mathematics Education.* Stanford, Calif.: School Mathematics Study Group, 1969, pp. 153-187.

In addition to this overview, you may want to read details of some exemplary studies. A classic work in the field is Duncker, Karl. "On Problem-Solving." *Psychological Monographs* 58 (Whole no. 270), 1945.

A specific heuristic described by Polya is studied in Anthony, W. S. "Working Backward and Working Forward in Problem Solving." *British Journal of Psychology* 57 (1966): 53-59.

A study which compares the problem-solving approaches of students with a computer program for translating English sentences to algebra equations is reported in Paige, Jeffrey M. and Herbert A. Simon. "Cognitive Processes in Solving Algebra Word Problems." Chap. 3. *Problem-Solving: Research, Method, and Theory.* Edited by Benjamin Kleinmuntz. New York: John Wiley & Sons, 1966.

The tendency to fix on a particular problem-solving approach even though it is inappropriate is discussed in Cunningham, John D. "Rigidity in Children's Problem Solving." *School Science and Mathematics* 66 (1966): 377-389.

Results of a program to teach problem-solving heuristics are reported in Covington, Martin V. and Richard S. Crutchfield. "Facilitation of Creative Problem Solving." *Programmed Instruction*, 4 (1965): 3-5, 10.

An instructional program aimed specifically at mathematics problem solving is reported in Riedesel, C. Alan. "Verbal Problem Solving: Suggestions for Improving Instruction." *The Arithmetic Teacher* 11 (1964): 312-316.

References to Unit 4

1. Butler, Charles H. and F. Lynwood Wren. *The Teaching of Secondary Mathematics.* 3rd ed. New York: McGraw-Hill Book Co., 1960.

2. Dodes, Irving Allen. "Planned Instruction." *The Learning of Mathematics: Its Theory and Practice.* Washington, D.C.: The National Council of Teachers of Mathematics, 1953.

3. Fawcett, Harold P. and Kenneth B. Cummins. *The Teaching of Mathematics From Counting to Calculus.* Columbus, Ohio: Charles E. Merrill Publishing Co., 1970.

4. Geleunter, H. L. and N. Rochester. "Intelligent Behavior in Problem-Solving Machines." *I.B.M. Journal of Research and Development* 2 (1958): 336-345.

5. Hartmann, George. "Gestalt Psychology and Mathematical Insight." *The Mathematics Teacher* 59 (1966): 656-661.

6. Henderson, Kenneth B. and Robert E. Pingry. "Problem Solving in Mathematics." *The Learning of Mathematics: Its Theory and Practice.* Washington, D.C.: National Council of Teachers of Mathematics, 1953.

7. Needleman, Joan R. "Discovery Approach — Polar Coordinates in Grade Seven?" *The Arithmetic Teacher* 14 (1967): 563-565.

8. Polya, George. *How to Solve It.* Garden City, N. Y.: Doubleday Anchor Books, 1957.

Unit 5

Concept Learning and Cognitive Structure

Facts, Concepts, and Principles

Gestalt psychologists are particularly concerned with patterns — with relationships between figure and ground. Mathematicians, of course, are also interested in patterns. The English-educated mathematician, W. W. Sawyer, goes so far as to define mathematics in terms of patterns. In his book, *Prelude to Mathematics*, he writes,

For the purposes of this book we may say, "Mathematics is the classification and study of all possible patterns." Pattern is here used in a way that not everybody may agree with. It is to be understood, in a very wide sense, to cover almost *any kind of regularity that can be recognized by the mind*. Life, and certainly intellectual life, is only possible because there are certain regularities in the world. A bird recognizes the black and yellow bands of a wasp; man recognizes that the growth of a plant follows the sowing of seed. In each case, a mind is aware of pattern.

Pattern is the only relatively stable thing in a changing world. Today is never exactly like yesterday. We never see a face twice from exactly the same angle. Recognition is possible not because experience ever repeats itself, but because in all the flux of life certain patterns remain identifiable. Such an enduring pattern is implied when we speak of "my bicycle" or "the river Thames," not withstanding the fact that the bicycle is rapidly rusting away and the river perpetually emptying itself into the sea.

Any theory of mathematics must account for both the power of mathematics, its numerous applications to natural science, and the beauty of mathematics, the fascination it has for the mind. Our definition seems to do both. All science depends upon regularities in nature; the classification of types of regularity, of patterns, should then be of practical value. And the mind should find pleasure in such a study. In nature, necessity and desire are always linked. If response to pattern is characteristic alike of animal and human life, we should expect to find pleasure associated with the response to pattern as it is with hunger or sex. (5:12)

This search for "all possible patterns" certainly seems like a large order, yet mathematics itself is a large order! We dare not limit ourselves to patterns involving quantities or patterns involving shape, or patterns involving sequence. We cannot restrict mathematics to an abstraction of only those things seen in the physical world, nor can we confine it solely to the abstract world of ideas. Perhaps the study of patterns is not too large a categorization after all. Sawyer goes ahead to elaborate that by pattern he means "almost any regularity that can be recognized by the mind." (5:12)

This sweeping inclusiveness provides mathematics with its enormous range of usefulness.

How do we study patterns? First of all we *classify* them. In the course of this clarification process, the *concepts* basic to mathematics are formed. Consider the concept "five." How does a child form this concept? By looking upon it as a way to classify a particular kind of pattern he sees in his world. "Five" is what is common between the fingers on each hand and the toes on each foot, between the tires of his father's car (including the spare in the trunk) and the wires in his mother's pastry blender. A concept is an idea which is formed by considering the common property of an appropriate set of exemplars. We shall take the existence of a set of exemplars as a criterion for identifying concepts. If we can find *more than one* example of an idea, we shall term it a *concept*. There are some ideas which have only one manifestation, and we shall term these ideas *facts*. Facts are more primitive than concepts. For example, the truth of the statement "$\frac{1}{3} < \frac{1}{2}$" is a fact under these conventions. One can give many plausible arguments to show why $\frac{1}{3}$ must be less than $\frac{1}{2}$, but one cannot give another example of this fact. In contrast, the statement "$x > y$ implies that $1/x < 1/y$" *is* a concept since we can call into evidence such facts as $\frac{1}{3} < \frac{1}{2}$, $\frac{1}{4} < \frac{1}{3}$, $\frac{1}{5} < \frac{1}{4}$, etc. as examples of the concept. Notice the power of this concept. It brings a kind of structure to the maze of fractional inequalities. It allows us to apply a test to determine the factual truth of statements like $\frac{1}{2439} < \frac{1}{2438}$. And it relates a subset of the rational numbers (those numbers of the form $1/n$) to the set of natural numbers.

Let us consider another example. Most people refer casually to addition and multiplication *facts*. Is a statement like "$2 + 3 = 5$" a fact or a concept under our conventions? In its most basic form, "$2 + 3 = 5$" is a single entry in an addition table. Because we cannot give other examples of correct entries for this particular position in the addition table, to consider it as a fact does seem correct. On the other hand, if we join a set of two apples with another (disjoint) set of three apples, the resulting set contains five apples. We can repeat this action for sets of sticks, stones, bones, toys, books, crayons, for sets whose members have common identifying characteristics, and for sets whose members have mixed identifying characteristics. Do these serve as examples of "$2 + 3 = 5$" and qualify it as a concept? In a technical sense they do not, for the operation of addition and the union of sets are not exactly the same thing. Addition is a binary operation which maps two numbers onto a single number. This mapping is specified by simply listing all pairs of numbers together with the single number they are mapped onto. We could write the list something like this:

Addition

$$(0,1) \longrightarrow 1$$
$$(1,1) \longrightarrow 2$$
$$(1,2) \longrightarrow 3$$
$$(2,2) \longrightarrow 4$$
$$(1,3) \longrightarrow 4$$
$$(2,3) \longrightarrow 5$$
$$(3,3) \longrightarrow 6$$
$$(1,4) \longrightarrow 5$$
$$(2,4) \longrightarrow 6$$
$$(3,4) \longrightarrow 7$$
$$(4,4) \longrightarrow 8$$

In practice, however, it is more convenient to use a two-dimension matrix to specify these mappings like the table below:

+	0	2	2	3	4
0	0	1	2	3	4
1	1	2	3	4	5
2	2	3	4	5	6
3	3	4	5	6	7
4	4	5	6	7	8

Here we find the number a particular pair is mapped onto by locating the row which contains the first number of the pair as its left-most element and the column which contains the second number of the pair as its top element. The intersection of that row with that column contains the number which the pair is mapped onto under the operation of addition.

When we define an operation we are essentially free to define it any way we want. If no one else had defined addition, we could map 2 and 3 onto any number we wanted. If we are the first to define the operation and want to map 2 and 3 onto the number 23, then for our "addition," the statement $2+3=23$ becomes a fact.

Of course, some definitions might be better than others. We would generally agree that a particular definition is better if it is more useful, that is, if it can be applied to a variety of other situations. The joining of two disjoint sets provides one such variety of useful situations. With a little care we can define addition in such a way that it corresponds to the number of members in two disjoint sets before and after they have been combined. If this definition is made, the mapping we call addition becomes more than just a game or memory task. It can be *applied* to describe and predict a certain characteristic of set union. The examples, "two stones joined with three stones result in a set of five stones," and "a set of two apples joined with a set of three apples result in a set of five apples," do not illustrate the addition fact, "2 + 3 = 5." They do illustrate the idea that this addition fact can be applied to the joining of certain sets of objects. This idea is a very powerful and relatively complex one. In particular, the application of the addition fact, "2 + 3 = 5," to the joining of certain sets is an idea which encompasses and includes several concepts. It relates the concept of addition of numbers to the concept of union of sets. Perhaps an idea as complex and powerful as this deserves the creation of another category. We shall call an idea which relates two or more concepts, a *principle*. Mathematics abounds with principles. "Multiplication is distributed over addition" certainly qualifies as a principle since it links the concepts of multiplication and addition. "Two distinct straight lines cannot intersect in more than one point" is a principle relating concepts of distinct straight lines, intersection, and point.

Mathematics, then, is a fabric woven of facts, concepts, and principles about patterns. Unlike most fabrics, the threads of mathematics — those facts, concepts, and principles — are not easy to trace or unravel. Is the idea of a commutative operation a concept or a principle? It involves ideas of operation, order, and elements of a set. But it does not explicitly establish any *relationship* between them. The notion of commutative operations seems to be a concept. On the other hand, the statement, "addition is commutative," does connect or relate the concept of a commutative operation with the concept of addition. This distinction is admittedly a fine one. Perhaps it would be helpful to note that in most cases mathematicians have given concepts names — either single words or short phrases. *Number, addition, mapping, function, equality, perpendicular lines, right triangles, interior,* and *limit* are all concepts. On the other hand, naming a principle usually requires a complete English sentence (although it may be a short sentence). "The sum of two odd numbers is an even number." "Natural numbers are not closed for subtraction." "Multiplication is associative." The statements are principles. Notice that they do not describe or elaborate an idea but simply relate or connect two or more ideas.

How do we learn concepts and principles? Principles are relationships and connections between concepts. That we cannot learn principles without previously having learned their component concepts seems reasonable. We cannot grasp the principle that addition is applicable to the union of disjoint sets without having learned at least on an intuitive level the concepts of union and set. Therefore we should begin by considering concept learning. Many mathematical concepts cannot easily be communicated in concise verbal expressions and may not even be capable of precise definition. The concept of "set" is an example. How do we learn then what is meant by a set? The answer, of course, is by abstracting the common property of a series of examples, which is exactly what we mean by a concept. As teachers, we should be aware that the simple giving of a definition is never sufficient for teaching a concept. The key to concepts is examples. We learn concepts by examining examples. Fortunately, this process is not limited to formal teaching situations. The ability to compare must depend upon the ability to discriminate, and we know that very young infants can discriminate between different stimuli. So the child learns concepts by comparing examples long before he comes to school. In a sense, we do not have to worry about how this process begins, for it will begin in spite of us. When it comes to learning, children are self-starters. What we do determines not so much whether learning will start, but whether or not it will continue.

Examples and Nonexamples in Concept Learning

We learn concepts by identifying the common characteristic of a series of examples. At the same time, however, to introduce examples which do not have the concept characteristic often seems appropriate. We shall refer to these as nonexamples of the concept. For instance, consider the concept of a commutative operation. Part of the depth of understanding of this concept comes from an understanding of operations which are *not* commutative as well as from those which are. Addition and multiplication are examples of the commutative concept, but subtraction and division are nonexamples. In this case it seems logical that nonexamples play just as important a role in concept formation as examples.

Eighty-one attribute cards are shown on page 196. Notice that these cards can be classified in many ways. We can consider all cards which have a single border, or a double border, or a triple border. We can consider all cards which contain crosses, or circles, or squares. We can consider cards in which the figures are white, or black, or cross-hatched. Or we can classify according to combinations of these attributes — for example, all cards which contain two crosses, or two circles, or two squares. For this experiment we will consider any such classification scheme to be a *concept*.

Suppose we selected a concept and ask a subject to identify it. He can identify by selecting cards one at a time and asking whether or not the card is an example of the concept the experimenter has in mind. We can reduce the random nature of this process by first showing a card which is an example of the concept. Since there are many different ways a single card could be classified, the subject must eliminate possibilities by comparing it with cards which are examples or nonexamples of the concept to be identified. Will he perceive that he has received as much information when he learns that he has selected a nonexample as when he has selected an example? We can explore this perception by asking him to predict whether his selection will be an example or a nonexample and to write down how much he would "bet" on his prediction. This symbolic "bet" can be taken as an indication of how sure he feels about his information. By looking at the size of the increase of the bet after correctly predicted example and comparing it with the increase of the size of the bet after a correctly predicted nonexample, we can begin to see if he perceives any differences between the information conveyed by examples and nonexamples.

We want to test this hypothesis: there will be no difference between the increase in the bet which follows a successful prediction of an example and the increase in the bet which follows a successful prediction of a nonexample.

For this experiment, we will show the subject the sheet of attribute cards on page 196, and ask him to record his responses himself on a copy of the response sheet on page 197. When the subject is his own recorder, he can refer back to previous choices and outcomes with a minimum of confusion. For this experiment we will use the concept "any two figures." Any card which contains two figures, whether they are crosses, circles, or squares, or whether they are white, black, or cross-hatched, will then be examples of this concept. Use the following statement as a guide for telling the subject what to do.

This is an experiment in which you will try to discover a concept I have in mind about the pictures on this display sheet. (Show the display sheet.) There are eighty-one cards on this sheet. All of them are different, but they do share common characteristics such as the number of borders, the shape of the central figures, or shading of these figures. The concept I will have in mind will be some particular way of grouping or combining these cards according to one or more characteristics. For example, if the concept I choose were "black squares," then cards like numbers 46, 47, 51, and 54 would belong to the group and be examples of the concept. A card like number 23 would be a nonexample of the concept since it does not contain a black square. Do you understand? For practice, why don't you make up a concept and show me some cards which are examples of this concept and some cards which are nonexamples of this concept. (Give the subject time to do this.)

Good! Now I'm going to think of a concept which I want you to guess. I will keep this same concept in mind for the rest of this experiment. To get you started I will give you the number of a card which is an example of this concept. You can write that number on this response sheet (hand the subject the response sheet). Then I want you to jot down your first guess about what you think the concept is. When you have done this I want you to choose a card, predict whether it is an example or nonexample of the concept, and write down how much you would "bet" that your prediction was right. We will not actually bet any money, but what you write down will give me an idea of how certain or uncertain you are about your prediction. When you have written all this down, I will tell you whether your prediction was right or wrong.

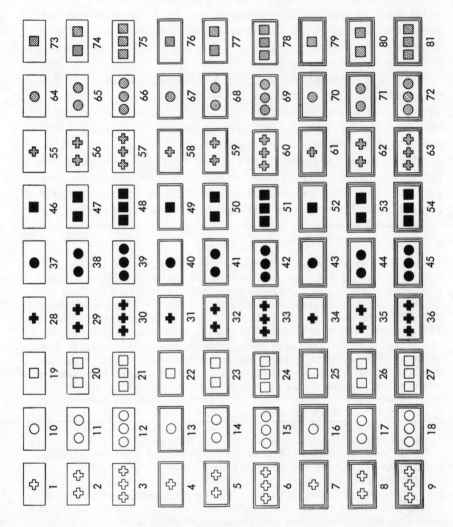

Figure 34. Card Display Sheet

Response Sheet:

Name: _____

Number of example card: _____

Trial Number	Guess About Concept	Card Choice (Number)	Prediction (Example or Nonexample)	"Bet" ($1 to $10)	Outcome (Correct or Incorrect)
1.					
2.					
3.					
4.					
5.					
6.					
7.					
8.					
9.					
10.					
11.					
12.					

Figure 35. Concept Response Sheet

Then you can make another guess or keep the same guess and repeat the process. Do you understand the procedure?" (Give the subject time to respond and answer any questions.)

Now there is one more thing I should tell you before we begin. I will tell you whether your prediction is right or wrong each time, but I will *not* tell you whether your guess about the concept is right or wrong until we have finished all twelve trials. So remember that we will not stop when you have guessed the right concept but will keep on going until twelve trials have been done. Since there are eighty-one cards you should not run out of examples or nonexamples to pick. Let's begin. An example of the concept I have in mind is card number 59.

When you have finished, tell the subject what the concept was and thank him for his cooperation. Repeat the experiment with at least three subjects.

Complete the following table from the response sheet of each of your subjects. Discard all trials in which the subject made the maximum "bet" of $10.

Trial Number of Correct Prediction	Amount Bet on This Trial ($10)	Amount Bet on Next Trial	Increase of Bet	Was an example or a nonexample predicted on this trial?

Table 11. Subject Responses.

We want to see if there is any difference between the increase of the bet following a correct prediction of an example and the increase of the bet following a correct prediction of a nonexample. Of course, there may be other factors which affected the increase of the bet besides the distinction between examples and nonexamples. One of the most important of these is the number of the trial. One might expect that the size of the bet would normally increase as the experiment progresses because of the cumulative effect of each piece of knowledge. On the other hand, this tendency may be offset by the fact that we imposed a ceiling or maximum size for the bet. It is conceivable that subjects might perceive the ceiling as a limiting factor and make somewhat smaller increases as they got closer and closer to the ten dollar maximum. Whether or not these two effects offset each other is something we cannot generally predict, but we must check for each set of data we want to consider. The easiest way to check these trends is to plot a scatter diagram of the amount of increase against the trial number as you have recorded from the summary table. Pool the data gathered by the entire class and plot it all on a single coordinate system. (See investigation 1-1 to review scatter diagrams.)

If the points of the scatter diagram appear to be randomly distributed throughout the graph quadrant, we can safely assume that the two trends have effectively offset each other. In this case, we can immediately begin to calculate whether or not there is a significant difference between the increase of the bet following the successful prediction of an example and the increase of the bet following the successful prediction of a nonexample. Consider one pair of an example and a nonexample for each subject. For some subjects you may find that they correctly predicted several examples and nonexamples before reaching the ten dollar limit. In this case, decide which pair to take by the following criterion: choose the example-nonexample pair that is separated by the fewest number of trials; if a tie exists, select which of the tied pairs will be used at random by tossing a coin.

Scatter diagram points clustered along a line of either positive or negative slope indicate correlation between the size of the increase and the number of the trial. Then considering only those subjects who made successful example and nonexample predictions which were not separated by more than one intervening trial will be necessary. This determination will reduce the size of effect of the trial number. Next, we must be sure that not all of these pairs are in the same direction; for example, the correctly predicted example does not *always* occur on the early trial and the nonexample on the later trial. The number of subjects for which the example prediction occurred first should equal the number of subjects for which the nonexample prediction occurred first. Subject selection can be done by repeating the experiment until the proper number of suitable subjects is obtained or by removing subjects in the larger group. If subjects are removed, they must be selected for removal at random—perhaps by drawing slips of paper containing subject's number from a hat.

Number each subject in the data pool (you should have at least ten subjects) and complete the following table. Be careful to always compute the difference in the same direction (example less nonexample) and include the proper positive or negative sign.

Subject Number	Increase Following Example (E)	Increase Following Nonexample (N)	Difference D = E - N

Table 12. Difference in Increasing Bet Following Examples and Nonexamples.

Test for statistical significance of this difference by computing the *t*-ratio:

$$t = \frac{\bar{D}}{\sqrt{\dfrac{\sum_{i=1}^{n}(D_i - \bar{D})^2}{N(N-1)}}}$$

and finding the corresponding probability value. (See the analysis of experiment 3-1 for a review of this procedure.)

Analyzing the Data

You were probably convinced while in the process of doing experiment 5-1 that the knowledge of both examples and nonexamples was equally useful in identifying a concept. Despite this conclusion, finding isolated instances of people who seemingly do not process both kinds of information in the same way is not unusual. These people simply tend to ignore cases of nonexamples and concentrate almost exclusively on only those cases which were identified as examples. Did you find any subjects who seemed to behave this way? Perhaps it occurred to you that these people could be taught to form concepts more efficiently.

This intriguing possibility has led to a wide variety of psychological studies devoted to the strategies different people employ in forming concepts. Some of the most interesting of these are described in the book, *A Study of Thinking*, by Jerome Bruner, Jacqueline Goodnow, and G. A. Austin. (4) Bruner and his associates have not only been able to identify different concept formation strategies, but have also determined that the efficiency of these strategies depends upon the type of concept which is to be formed. As a result, Bruner has advocated that for school subjects such as mathematics, we should not limit ourselves to the teaching of mathematical concepts, but should also include the teaching of the processes that are used by mathematicians in forming these concepts. The key to teaching processes is, for Bruner, discovery teaching.

One cognitive psychologist who believes that Bruner has been excessive in emphasizing discovery is David Ausubel. (3) Ausubel holds that most of our concepts are formed on the basis of what others tell us. He feels that we should not ignore this reception learning, but instead, we should ask ourselves why some reception learning tends to efficiently foster concept formation while other reception learning produces only memorized responses which are quickly forgotten. The key to this distinction is, for Ausubel, contained in the notion of meaningfulness. The following reading module contains an article by Ausubel which discusses some of the factors in teaching which influence the meaningful formation of concepts.

Facilitating Meaningful Verbal Learning in the Classroom

David P. Ausubel

In mathematics, as in other scholarly disciplines, pupils acquire subject-matter knowledge largely through meaningful reception learning of presented concepts, principles, and factual information. In this paper, therefore, I first propose to distinguish briefly between reception and discovery learning, on the one hand, and between meaningful and rote learning, on the other. This will lead to a more extended discussion of the nature of meaningful verbal learning (an advanced form of meaningful reception learning) and the reasons it is predominant in the acquisition of subject matter; of the manipulable variables that influence its efficiency; and of some of the hazards connected with its use in the classroom setting.

Reception Versus Discovery Learning

The distinction between reception and discovery learning is not difficult to understand. In reception learning the principal content of what is to be learned is presented to the learner in more or less final form. The learning does not involve any discovery on his part. He is required only to internalize the material or incorporate it into his cognitive structure so that it is available for reproduction or other use at some future date. The essential feature of discovery learning, on the other hand, is that the principal content of what is to be learned is not given but must be discovered by the learner before he can internalize it; the distinctive and prior learning task, in other words, is to discover something. After this phase is completed, the discovered content is internalized just as in reception learning.

Meaningful Versus Rote Learning

Now this distinction between reception and discovery learning is so self-evident that it would be entirely unnecessary to belabor the point if it were not for the widespread but unwarranted belief that reception learning is invariably rote, and that discovery learning is invariably meaningful. Actually, each distinction constitutes an entirely inde-

Reprinted from *The Arithmetic Teacher,* vol. 15 (February 1968) 126-32. © 1968 by the National Council of Teachers of Mathematics. Used by permission.

pendent dimension of learning. Thus reception and discovery learning can each be rote or meaningful, depending on the conditions under which learning occurs. In *both* instances meaningful learning takes place if the learning task is related in a nonarbitrary and nonverbatim fashion to the learner's existing structure of knowledge. This presupposes 1) that the learner manifests a *meaningful learning set*, that is, a disposition to relate the new learning task nonarbitrarily and substantively to what he already knows, and 2) that the *learning task is potentially meaningful* to him, namely, relatable to his structure of knowledge on a nonarbitrary and nonverbatim basis. The first criterion, nonarbitrariness, implies some plausible or reasonable basis for establishing the relationship between the new material and existing relevant ideas in cognitive structure. The second criterion, substantiveness or nonverbatimness, implies that the potential meaningfulness of the material is never dependent on the exclusive use of particular words and no others, i.e., that the same concept or proposition expressed in synonymous language would induce substantially the same meaning.

The significance of meaningful learning for acquiring and retaining large bodies of subject matter becomes strikingly evident when we consider that human beings, unlike computers, can incorporate only very limited amounts of arbitrary and verbatim material, and also that they can retain such material only over very short intervals of time unless it is greatly overlearned and frequently reproduced. Hence, the tremendous efficiency of meaningful learning as an information-processing and -storing mechanism can be largely attributed to the two properties that make learning-material potentially meaningful.

First, by nonarbitrarily relating potentially meaningful material to established ideas in his cognitive structure, the learner can effectively exploit· his existing knowledge as an ideational and organizational matrix for the understanding, incorporation, and fixation of new knowledge. Nonarbitrary incorporation of a learning task into relevant portions of cognitive structure, so that new meanings are acquired, also implies that the newly learned meanings become an integral part of an established ideational system; and because this type of anchorage to cognitive structure is possible, learning and retention are no longer dependent on the frail human capacity for acquiring and retaining arbitrary associations. This anchoring process also protects the newly incorporated material from the interfering effects of previously learned and subsequently encountered similar materials that are so damaging in rote learning. The temporal span of retention is therefore greatly extended.

Second, the substantive or nonverbatim nature of thus relating new material to and incorporating it within cognitive structure circumvents the drastic limitations imposed by the short item and time spans of verbatim learning on the processing and storing of information. Much more can obviously be apprehended and retained if the learner is required to assimilate only the substance of ideas rather than the verbatim language used in expressing them.

It is only when we realize that meaningful learning presupposes only the two aforementioned conditions, and that the rote-meaningful and reception-discovery dimensions of learning are entirely separate, that we can appreciate the important role of meaningful reception learning in classroom learning. Although, for various reasons, rote reception learning of subject matter is all too common at all academic levels, this need not be the case if expository teaching is properly conducted. We are gradually beginning to realize not only that good expository teaching can lead to meaningful reception learning but also that discovery learning or problem solving is no panacea that guarantees meaningful learning. Problem solving in the classroom can be just as rote a process as the outright memorization of a mathematical formula without understanding the meaning of its component terms or their relationships to each other. This is obviously the case, for example, when students simply memorize rotely the sequence of steps involved in solving each of the "type problems" in a course such as algebra (without having the faintest idea of what they are doing and why) and then apply these steps mechanically to the solution of a given problem, after using various rotely memorized cues to identify it as an exemplar of the problem type in question. They get the right answers and undoubtedly engage in discovery learning. But is this learning any more meaningful than the rote memorization of a geometrical theorem as an arbitrary series of connected words?

In meaningful classroom learning, the balance between reception and discovery learning tends, for several reasons, to be weighted on the reception side: First, because of its inordinate time-cost, discovery learning is generally unfeasible as a *primary* means of acquiring large bodies of subject-matter knowledge. The very fact that the accumulated discoveries of millennia can be transmitted to each new generation in the course of childhood and youth is possible only because it is so much less time-consuming for teachers to communicate and explain an idea meaningfully to pupils than to have them rediscover it by themselves. Second, although the extent and complexity of meaningful reception learning in pure verbal form is seriously limited in pupils who are either cognitively immature in general or unsophisticated in a particular discipline, the actual process of discovery per se is never required for the meaningful acquisition of knowledge. Typically it is more efficient pedagogy to compensate for such deficiencies by simply incorporating concrete-empirical props into expository teaching techniques. Finally, although the development of problem-solving ability as an end in itself is a legitimate objective of education, it is less central an objective than that of learning the subject matter. The ability to solve problems calls for traits such as flexibility, originality, resourcefulness, and problem-sensitivity that are not only less generously distributed in the population of learners than is the ability to understand and retain verbally presented ideas but are also less teachable. Thus relatively few good problem solvers can be trained in comparison with the number of persons who can acquire a meaningful grasp of various subject-matter fields.

The Nature of Meaningful Reception Learning

Like all learning, reception learning is meaningful when the learning task is related in nonarbitrary and nonverbatim fashion to relevant aspects of what the learner already knows. It follows, therefore, from what was stated above that the first precondition for meaningful reception learning is that it take place under the auspices of a meaningful learning set. Thus irrespective of how much potential meaning may inhere in a given proposition, if the learner's intention is to internalize it as an arbitrary and verbatim series of words, both the learning process and the learning outcome must be rote or meaningless.

One reason why pupils commonly develop a rote-learning set in relation to potentially meaningful subject matter is that they learn from sad experience that substantively correct answers lacking in verbatim correspondence to what they have been taught receive no credit whatsoever from certain teachers. Another reason is that because of a generally high level of anxiety or because of chronic failure experience in a given subject (reflective, in turn, of low aptitude or poor teaching), they lack confidence in their ability to learn meaningfully, and hence they perceive no alternative to panic apart from rote learning. This phenomenon is very familiar to mathematics teachers because of the widespread prevalence of "number shock" or "number anxiety." Lastly, pupils may develop a rote-learning set if they are under excessive pressure to exhibit glibness, or to conceal rather than admit and gradually remedy original lack of genuine understanding. Under these circumstances it seems both easier and more important to create a spurious impression of facile comprehension by rotely memorizing a few key terms or sentences than to try to understand what they mean. Teachers frequently forget that pupils become very adept at using abstract terms with apparent appropriateness — when they have to — even though their understanding of the underlying concepts is virtually nonexistent.

The second precondition for meaningful reception learning—that the learning task be potentially meaningful or nonarbitrarily and substantively relatable to the learner's structure of knowledge—is a somewhat more complex matter than meaningful learning set. At the very least it depends on the two factors involved in establishing this kind of relationship, that is, on the nature of the material to be learned and on the availability and other properties of relevant content in the particular learner's cognitive structure. Turning first to the nature of the material, it must obviously be sufficiently plausible and reasonable that it could be related on a nonarbitrary and substantive basis to *any* hypothetical cognitive structure exhibiting the necessary ideational background. This is seldom a problem in school learning, since most subject-matter content unquestionably meets these specifications. But inasmuch as meaningful learning or the acquisition of meanings takes place in *particular* human beings, it is not sufficient that the learning task be relatable to relevant ideas simply in the abstract sense of the term. It is also necessary that the cognitive structure of the *particular* learner include relevant ideational content to which the learning task can be related. Thus, insofar as meaningful learning outcomes in the classroom are concerned, various properties of the learner's cognitive structure constitute the most crucial and variable determinants of potential meaningfulness. These properties will be considered briefly but systematically in the following section.

At this point it is important to appreciate that the idiosyncratic nature of each learner's cognitive structure implies that the meanings he acquires from any potentially meaningful learning task must necessarily be idiosyncratic in nature. In fact, it could hardly be an overstatement of the case to say that the extent to which learning is meaningful largely depends on how idiosyncratic it is — that is, on how intimately the objective content of the learning task can be incorporated into the distinctively idiosyncratic aspects of the learner's relevant cognitive structure. Thus in a very real sense the meaningfulness of reception learning is in large measure a function of how actively and energetically a given pupil endeavors to translate new propositions into terminology consistent with his particular vocabulary and ideational background, and how self-critical he is in judging whether this goal has been accomplished. The main danger relative to meaningful reception learning is not so much that the learner will frankly adopt a rote-learning set but that he will be insufficiently energetic in reformulating presented propositions so that they have real meaning for him in terms of his own structure of knowledge, and that he will then delude himself and his teachers into believing that the resulting empty, vague, or imprecise verbalisms are genuinely meaningful.

The Role of Cognitive Structure Variables in Meaningful Verbal Learning

Since, as suggested above, the potential meaningfulness of a learning task depends on its relatability to a particular learner's structure of knowledge in a given subject-matter area or subarea, it follows that *cognitive structure itself*, that is, both its substantive content and its major organizational properties, should be the principal factor influencing meaningful reception learning and retention in a classroom setting. According to this reasoning, it is largely by strengthening salient aspects of cognitive structure in the course of prior learning that new subject-matter learning can be facilitated. In principle, such deliberate manipulation of crucial cognitive structure variables — by shaping the content and arrangement of antecedent learning experience — should not meet with undue difficulty. It could be accomplished 1) *substantively*, by using for organizational and integrative purposes those unifying concepts and principles in a given discipline that have the greatest inclusiveness, generalizability, and explanatory power, and 2) *programmatically*, by employing optimally effective methods of ordering the sequence of subject matter, constructing its internal logic and organization, and arranging practice trials.

Both for research and for practical pedagogic purposes it is important to identify those manipulable properties or variables of existing cognitive structure that influence the meaningful reception learning of subject-matter knowledge. On logical grounds, three such variables seem self-evidently significant: 1) the *availability* in the learner's cognitive structure of relevant and otherwise appropriate ideas to which the new learning material can be nonarbitrarily and substantively related, so as to provide the kind of anchorage necessary for the incorporation and long-term retention of subject matter; 2) the extent to which such relevant ideas are *discriminable* from similar new ideas to be learned so that the latter can be incorporated and retained as separately identifiable entities in their own right; and 3) the *stability* and *clarity* of relevant anchoring ideas in cognitive structure, which affect both the strength of the anchorage they provide for new learning material and their degree of discriminability from similar new ideas in the learning task.

Availability of Relevant Anchoring Ideas in Cognitive Structure

One of the principal reasons for rote or inadequately meaningful learning of subject matter is that pupils are frequently required to learn the specifics of an unfamiliar discipline before they have acquired an adequate foundation of relevant and otherwise appropriate anchoring ideas. Because of the unavailability of such ideas in cognitive structure to which the specifics can be nonarbitrarily and substantively related, the latter material tends to lack potential meaningfulness. But this difficulty can largely be avoided if the more general and inclusive ideas of the discipline, that is, those which typically have the most explanatory potential, are presented first and are then progressively differentiated in terms of detail and specificity. In other words, meaningful reception learning and retention occur most readily and efficiently if, by virtue of prior learning, general and inclusive ideas are already available in cognitive structure to play a *subsuming* role relative to the more differentiated learning material that follows. This is the case because such subsuming ideas when established in the learner's structure of knowledge 1) have maximally specific and direct relevance for subsequent learning tasks, 2) possess enough explanatory power to render otherwise arbitrary factual detail potentially meaningful (i.e., relatable to cognitive structure on a nonarbitrary basis), 3) possess sufficient inherent stability to provide the firmest type of anchorage for detailed learning material, and 4) organize related new facts around a common theme, thereby integrating the component elements of new knowledge both with each other and with existing knowledge.

One of the more effective strategies that can be used for implementing the principle of progressive differentiation in the arrangement of subject-matter content involves the use of special introductory materials called "organizers." A given organizer is introduced in advance of the new learning task per se; is formulated in terms that, among other things, relate it to and take account of generally relevant background ideas already established in cognitive structure; and is presented at an appropriate level of abstraction, generality, and inclusiveness to provide specifically relevant ideational scaffolding for the more differentiated and detailed material that is subsequently presented. An additional advantage of the organizer, besides guaranteeing the availability of specifically relevant anchoring ideas in cognitive structure, is that it makes explicit both its own relevance and that of the aforementioned background ideas for the new learning material. This is important because the mere availability of relevant anchoring ideas in cognitive structure does not assure the potential meaningfulness of a learning task unless this relevance is appreciated by the learner. Lastly, it is desirable not only for the material within each topic to become progressively more differentiated — both by using organizers and by proceeding from subtopics of greater to lesser inclusiveness in the learning material itself — but also to follow the same organizational plan in ordering the sequence of the various topics comprising a given course of study.

It is also possible in subject-matter learning to capitalize on the availability in cognitive structure of relevant anchoring ideas reflective of prior incidental experience or nonverbal learning. This is the underlying rationale for the widely accepted pedagogic practice of proceeding from intuitively familiar to intuitively unfamiliar topics in sequencing subject matter, thereby using previously acquired intuitive principles or general background as a foundation for learning less familiar material.

Finally, the availability of relevant anchoring ideas for use in meaningful verbal learning may be maximized by taking advantage of natural sequential dependencies among the component divisions of a particular discipline, i.e., of the fact that the understanding of a given topic often logically presupposes the prior understanding of some related topic. Thus, by arranging the order of topics and subtopics in a given subject-matter field as far as possible in accordance with these sequential dependencies, the learning of each unit, in turn, not only becomes an achievement in its own right but also constitutes specifically relevant ideational scaffolding for the next item in the sequence.

Consolidation of Anchoring Ideas

The sequential organization of subject matter naturally assumes that any given step in a particular sequence is always clear and stable before the next step is presented. If this is not the case, the anchorage it furnishes for all subsequent steps is insecure, and their learning and retention are accordingly jeopardized. Hence, new material in the sequence should never be introduced until the preceding step is thoroughly mastered. Such mastery, of course, can be achieved only through adequate and differential practice, review, testing, and feedback which provide the necessary confirmation, clarification, and correction required for the effective consolidation of meaningfully learned material. Consolidation also facilitates meaningful verbal learning by increasing the discriminability of previously learned material from similar new learning tasks—those that are sequentially dependent and those that are not.

Discriminability of Learning Material from Established Ideas

This brings us to the role of discriminability in meaningful verbal learning. It is self-evident that before new ideas can be meaningfully learned, they must be adequately discriminable from similar established ideas in cognitive structure. If the learner cannot discriminate clearly, for example, between new idea A' and previously learned idea A, then A' enjoys relatively little status as a separately identifiable meaning in its own right, even at the very onset of its incorporation into cognitive structure. In addition, if new meanings cannot be readily distinguished from previously learned established meanings, they can certainly be adequately represented by them for memorial purposes, and thus they tend to be reduced to the latter even more rapidly than is typically the case in the retention of new meanings. In other words, only discriminable variants of established ideas in cognitive structure have long-term retention potentialities.

Thus, in learning situations where new ideas are introduced that are similar to previously learned ideas and hence confusable with them, it is advisable, by means of a procedure known as *integrative reconciliation*, to point out *explicitly* the basic similarities and differences between them. This practice integrates knowledge by specifically identifying the commonalities underlying similar ideas; by preventing artificial compartmentalization and the proliferation of separate terms for concepts that are basically identical except for contextual usage; and, most important, by sharply delineating in what ways similar but not identical ideas are actually different. Failure to specify such relationships between previously acquired and later-appearing subject-matter content, that is, treating the latter content in self-contained fashion without explicitly attempting to reconcile it with the former, assumes rather unrealistically that students will adequately perform the necessary cross-referencing.

Where necessary, organizers can also further the goal of integrative reconciliation by explicitly delineating the essential similarities and differences between the new subsuming concepts and principles to be learned and similar established ideas in cognitive structure. By so enhancing the discriminability of the newly introduced anchoring ideas, such organizers enable the learner to grasp the more differentiated aspects of the new learning task with many fewer ambiguities and misconceptions than would otherwise be possible. This differentiated material is also retained longer both because it is learned more clearly in the first place (by virtue of the greater discriminability of the new anchoring ideas under which it is subsumed) and because more discriminable subsumers are themselves more stable and hence better able to provide secure anchorage.

The Structure
of Mathematics

The distinction between facts and concepts and between concepts and principles is not always an easy distinction to make. But if we can make it for only a majority of cases, we can begin to gain an insight into meaningful learning. Ausubel holds that for meaningful learning to occur the learning task must be meaningful to the learner, "namely, relatable to his structure of knowledge on a nonarbitrary and nonverbatim basis." What we have developed is one way in which mathematics (and knowledge in general) is structured — in terms of facts, concepts, and principles. The *fact* "2+3=5" cannot possibly be meaningful unless it can be related to the *concept* of addition. And the *concept* of addition is meaningless until it can be related to some *principle*. Exactly what kind of principle is necessary depends upon the developmental stage or ability of the learner. At an advanced level of abstract operation (á la Piaget) it may be sufficient to relate addition to the *principle* that a binary operation is a mapping or function from a set of pairs of given elements onto a set of unitary elements. On the other hand, if this stage has not been achieved, this kind of concept-principle relationship is not available to the learner. We should look for relationships to principles that are available at lower stages. At a stage of concrete operations, we could expect to relate the *concept* of addition to the *principle* which connects addition to the union of disjoint sets, and indeed this is what is almost invariably done in the elementary school.

This does not mean that we cannot teach a child to say "2+3= 5" without relating it to concepts or principles. We can use operant conditioning to shape his response patterns appropriately, and finding a "precocious preschooler" who can respond "5" to the stimulus "2+3" because his parents have done just that is not too hard. But this kind of learning is rote learning. Ausubel is postulating that we cannot have meaningful learning, even of facts, until the facts can be joined into a knowledge structure of concepts and principles. These concepts and principles must have already been learned in some fashion by the learner. Since every individual is likely to have a slightly different set of concepts and principles at his disposal, Ausubel is inclined to talk about the cognitive structure of the learner, rather than the more theoretical cognitive structure of the subject matter.

In Ausubel's theory, the learning of ideas takes place when a new idea is assimilated into an existing cognitive structure. This existing cognitive structure is an all-important prerequisite. Ausubel says, "If I had to reduce all the educational psychology to just one principle, I would say this: The most important single factor influencing learning is what the learner already knows. Ascertain this and teach him accordingly." We should not delude ourselves into restricting this existing cognitive structure to what the student has been taught formally. The first grader on the first day of school brings cognitive structures which have been formed by five years of active explorations. To be sure, these structures are limited, fractional, and in some cases naive, but they exist. No learner operates out of an absolute vacuum.

One variable influencing learning and retention is the availability in cognitive structure of specifically relevant anchoring ideas, however. While the fact that the learner is without an existing cognitive structure is highly improbable, that there are no ideas within this structure which are relevant to the new ideas to be learned is entirely possible. If meaningful learning is to occur, this relevance gap must be bridged. The alternative is a regression to nonmeaningful or rote learning. One way to bridge this gap involves the use of appropriately relevant and inclusive introductory materials. Ausubel calls these *organizers*. Advance organizers show the structure of the concepts to be learned at a level of abstraction, generality, and inclusiveness which is much higher than that of the learning material which is to follow. In giving the learner a general structure prior to the presentation of detailed facts, such organizers can provide a framework of necessary anchoring ideas.

Anchoring ideas can operate to bring about subsumption of the new idea in two ways. When the new idea is a specific example of the anchoring idea, the subsumption is derivative. Although such material may serve to strengthen the anchoring idea, derivative ideas tend to be both quickly understood and quickly forgotten. (1:100) When the new idea is an elaboration or modification of the anchoring idea, its subsumption is correlative. In this case, the new material extends, elaborates, modifies, or qualifies previously learned ideas. New subject matter is more typically learned through the process of correlative subsumption than derivative subsumption. (1:100)

Cognitive Structure in the Classroom

The terminology that Ausubel uses — *cognitive structure, anchoring ideas, derivative and correlative subsumption, ideational scaffolding* — does not readily suggest classroom behaviors or teaching principles. What kind of evidence can we see in the behavior of students which would be consonant with Ausubel's analysis of the learning process? What kinds of teaching acts might be suggested by this theory?

Every mathematics teacher has seen evidence that children intuitively look for structure into which new ideas can be fitted, assimilated, or accommodated. This evidence usually takes the form of the plaintive questions, "Why do we have to learn this?" "What good is mathematics, anyway?" "What will I ever be able to use this for?" In more sophisticated terms, these questions become a demand for *relevance* in the curriculum. Unfortunately, we tend not to see these questions for what they really are — attempts to see how the new information being taught fits into what is already known or, even better, what is yet to come. We see these questions, instead, as challenges to our value system. As teachers we have accepted the particular piece of information we are teaching as being important, and our immediate reaction is that we must defend its importance. But judging the value of a piece of information requires looking at that information in retrospect, and such a position is one that students cannot possibly take. Frustrated, we retreat to a position of authority. We say, "You need to study mathematics to get into college." "You have to have mathematics to graduate from high school." But these statements are not really reasons when carefully examined. Establishing mathematics as a criteria for entering a system or for leaving a system is, after all, highly arbitrary and authoritarian. The kind of mathematics required for most jobs is highly specific and could well be learned without the broad, general background taught in most school mathematics courses.

Look again at the student's questions. "Why do I have to learn to multiply binomials?" This question does not really challenge the importance of bonomial multiplication as much as it asks, "How does it relate?" "Where will it fit?" "What is the *structure?*" But the answers the child usually gets are along the lines of, "Because it's in the textbook." "Because that's what algebra I is all about." "Because you have to know that to take algebra II." None of these answers really provides the sought-for structure. After several exchanges of this nature, usually early in the school year, the student submits to rotely learn the facts of mathematics.

As teachers of modern mathematics, we like to claim that we are teaching structure. In fact, an examination of most mathematics textbooks will reveal that the structure, though present, is lost in a bewildering maze of factual information, skills, and exercises. The structure of algebra, for example, is usually relegated to a single chapter which discusses the basic properties of the real numbers along the lines of Peano's postulates. This kind of cognitive structure is decidedly not what Ausubel is calling for.

But what is a teacher to do? How does one construct the ideational scaffolding that allows students to attach and anchor ideas? The key is the idea of continuity — not a mathematical continuity, but a continuity of ideas. This aspect is the important one of scaffolding. A scaffold is continuous, each member connecting securely to another member until the whole becomes stable. An ideational scaffold should be constructed the same way. Remove enough connections from a real scaffold and it collapses. Remove connections between ideas, and meaningful learning collapses in a similar manner.

Because mathematics is, by its very nature, sequential, we enjoy an ideal situation for stressing the continuity of mathematical ideas. Yet most texts do not stress continuity. A chapter on polynomials begins with a definition of a polynomial. Then, bang! we are off and running — adding, subtracting, multiplying, and dividing — with no explanation as to how these operations all fit together. Why do we study polynomials in algebra? Are they another example of a ring? Do they provide a way to classify algebraic equations? Or have we suffered through all those manipulative contortions and feel obligated to see that others suffer as much as we did? Unfortunately, we seldom relate the polynomial chapter to previous chapters, nor do we first look at future problems where the ability to manipulate polynomials would be very useful.

Consider a chapter involving the simplification of radical expressions. These chapters always precede the chapters on quadratic and higher degree equations. If we turned these chapters around and first considered quadratic equations, we would see that radicals occur very naturally, and that we should turn our attention to the processes for simplifying radical expressions would seem very plausible because a connection point — the solution sets of quadratic equations — allows the continuity of ideas. Of course, this method is not the only way to provide for continuity of these ideas. There are many ways to trace a continual flow of ideas through mathematics, just as there are many ways to trace a continuous path through the framing of a scaffold. If meaningful learning in mathematics is our only goal, *which* continual trace we choose is unimportant, that the trace is *continuous* is important. Reviewing past material and showing how new material forms similar patterns and structures provides continuation. We can often look into the future by choosing a problem which students do not have the skill to solve and discuss in general ways the skill areas which might be useful in solving the problem.

Structure and the Nature of Mathematics

We see that the cognitive structure of mathematics itself becomes the most important factor influencing meaningful learning. New subject matter learning is facilitated by strengthening the organizational properties of the content in the course of prior learning. What are these organizational properties of mathematics? Until we can come to grips with this question, the ultimate understanding of meaningful learning will elude us. The guiding question of past units, "How do we learn mathematics?" has led us to an even more fundamental question: "What is mathematics?"

Systems and Abstractions

Somewhere at the core of mathematics lies the basic notion of a *system*. One way to define mathematics is to consider it as a *collection of interrelated abstract symbolic systems*. To explore this definition, consider some examples of such systems. The most familiar system to most people is the system of whole numbers. The whole number system is made up of three basic kinds of sets. The first of these is the set of elements or digits:

$$A = \{0, 1, 2, 3, 4, 5, 6, 7, 8, 9\}$$

The second basic set is the set of operations — addition, subtraction, multiplication, and division. This set is conveniently represented by the following symbols:

$$B = \{+, -, \times, \div\}$$

Finally, we have a basic set of relations — equal, less than, more than, and their respective negations. These are symbolized as:

$$C = \{=, <, >, \neq, \not<, \not>\}$$

In practice we seldom work with any of these three basic kinds of elements by themselves. Instead we work with combinations of the elements. There are three fundamental types of combinations.

The first type is called a "word." A word is a finite horizontal sequence of elements contained in set A. Thus, 1234 is a word in our system, as is 827610593, or 499763112085. Of course any single element of set A can also be considered as a word since, for example, 2 is a finite horizontal sequence of these elements. As with language, there are orthographic rules governing the construction of these sequences. For the system of whole numbers, however, this orthography usually consists of only one rule: a sequence may not begin with the element "0." The words of our system are commonly referred to as "numbers," of course.

A second type of combination which may be considered is a "phrase." A phrase consists of two or more words connected by elements from the set of operations (B). Thus, $2+3$, $4 \div 2$, $6-1$, and 7×5 are simple phrases.

Finally we consider combinations called "sentences." A sentence is made up of two words, two phrases, or one phrase and one word connected by an element from the set of relations (C). Thus, $2+3=4+1$, $5<8$, and $7-5>1+0$ are sentences.

Now all sentences may be placed in one of two categories: They are either true or false. The system contains a variety of procedures or algorithms for determining the truth or falsity of a statement. For example, to determine if a sentence of the form $N+M=P$ is true or false (where M and N are elements of the set A), the algorithm refers one to the following matrix:

+	0	1	2	3	4	5	6	7	8	9
0	0	1	2	3	4	5	6	7	8	9
1	1	2	3	4	5	6	7	8	9	10
2	2	3	4	5	6	7	8	9	10	11
3	3	4	5	6	7	8	9	10	11	12

If M and N are words of more than one digit, the table is used in certain ways for parts of the word in a process known as a computational algorithm.

Just as an addition table is used to determine the truth or falsity of sentences of the form $N+M=P$, so a subtraction table is used as a criteria for determining the truth or falsity of sentences of the form $N-M=P$. In matrix form, a subtraction table contains an interesting property: some of the cells of the matrix are empty. This feature allows us to classify our operations into two groups. If the operation matrix does not contain empty cells, the operation is known as a *closed* one; if the operation matrix does contain empty cells, the operation is called *not closed*.

The truth of a sentence which relates two words can be determined by referring to the subtraction matrix. $N = M$ if the sentence $N - M = 0$ is true. $N < M$ is true if the phrase $M - N$ refers to a cell in the subtraction matrix which is not empty and does not contain 0. Similarly, $N > M$ is true if the phrase $N - M$ refers to a nonempty or nonzero cell.

There are some sentence patterns in the system which are always true for any words under the usual rules of substitutions. Thus, $N + M = M + N$ and $M \times N = N \times M$ are always true as is the sentence pattern $2 \times N - N = N$. These sentence patterns are often referred to as "laws" or "principles" of the system.

Although we have touched upon most of the basic features of the whole number system, we have by no means exhausted all of the requirements made upon the system or all of its properties. One way in which the system is enriched is by imposing definitions upon it. We may define what we mean by a prime number or a triangular number and study the properties of these subsets of the system.

The system we have just discussed is a mathematical, symbolic, abstract system. It is a mathematical system because it consists of a set of elements, a set of operations, and a set of relations with certain operating procedures or rules. It is *symbolic* because each of these sets is comprised of members which are symbols. The rules for combining these symbols into words, phrases, and sentences can be handled by a computer as well as by a child. Although the child may need to relate the symbols to concrete objects and motions to learn the system meaningfully, this restriction tells us something about the properties of children and not about any property of the system. Certainly a computer can handle the symbolic system without recourse to anything but the symbols.

What makes the system *abstract?* Does saying that a system is abstract mean any more than saying that the system is symbolic? To assume that every symbolic system is abstract seems reasonable. But is every abstract system symbolic? Let us look briefly at an abstract system which is not symbolic. We need a set of elements, a set of operations, and a set of relations, none of which are symbols. Consider as the set of elements any finite set of concrete, real objects. Imagine each book in your college library as an element of this basic set. Words in this system will be stacks of these elements — one book placed on top of another.

For the second basic set we need an operation. Take as an operation the physical placement of one book on top of another, or the physical placement of one stack of books upon another stack of books. We might call this operation "pile."

For our third basic set, we need to define a set of relations. We will define the relations *equal* and *smaller*, by specifying the following process. To determine which relation holds between two stacks, one must simultaneously remove one book from each of the stacks, repeating the process until the floor is reached. If the floor is reached first for one of the stacks it is smaller than the other stack. Sentences which involve phrases (the "pile" of two or more stacks) are judged by first forming single stacks according to the definition of "pile."

Now the system we have just discussed is not a symbolic system. Although we could represent the elements of our system with symbols, the system is not defined in terms of these symbols. Yet the system we have just described is *abstract*. None of the physical or material properties of the objects (books) are relevant to our system. We may have red books or blue books, thick books or thin books, heavy books or light books. In fact, part of the ability to successfully understand this system depends upon an ability to ignore physical properties. As we have defined "smaller," the smaller stack of books could be heavier than the second stack or even taller than the second stack.

All symbolic systems are also abstract for the same reason. As objects, all symbols have many properties which are irrelevant to the system. For example, some symbols are symmetric while others are asymmetric. Some symbols require more ink to print than others, some are simple closed curves, but in all cases, these properties are irrelevant to the system. All symbolic systems are abstract, but not all abstract systems are symbolic.

Finally, mathematics is a collection of *interrelated* abstract symbolic systems. This interrelatedness is, perhaps, what provides the ultimate power of mathematics. If we restrict the operation "pile" in our last example to mean the union of disjoint sets, then we can relate the system of concrete objects to the system of whole numbers under addition (only). The element of one system is a stack, the word in the other system is a number, and we can devise a correspondence (mapping) of each possible stack into the possible whole numbers. Furthermore, we can find a mapping so that the stack which results from the pile of two original stacks will map onto the number which results as the addition of those exact two numbers which the original stacks mapped onto. If we can do this mapping, we are freed of lugging about armloads of dusty books and piling them into stacks. Instead, we can simply manipulate symbols for we know that the addition of our symbols is mirrored in the piling of stacks of books. Thank goodness for lazy mathematicians! Their abstracting and generalizing often pays off in unexpected ways.

Geometric Systems

What we have just presented is a kind of advance organizer for the structures of mathematics. What we need to do now is fill in specific details. Our organizer has been generally described with special reference to the whole numbers of arithmetic. Can we find the same kind of structure in geometry? Of our three basic sets — elements, operations, and relations — nonmetric Euclidean geometry focuses most on the set of relations. The basic elements in this set are the congruent and similar relations, but such "properties" as perpendicular and parallel are also relations. The elements of the system are not only the familiar point, line, and plane but also line segments, rays, angles, and the combinations known as geometric figures.

Geometric figures, which are, in essence, only points combined by intuitively simple orthographic conventions, are the words of the system. Although there are operations in the system (angles can be combined, for example), the importance of using operations to combine words into phrases is relatively unimportant. Almost all sentences consist of two words separated by a relation. The definitions of Euclidean geometry establish the criteria for judging whether a sentence is true or false, and the postulates allow us to search for equivalent criteria.

If we change the geometric system, we can change the relative importance of each of the three basic sets. In particular, there are several ways we can increase the role of the set of operations in geometry. We can introduce numbers into the system by means of measurement postulates (often called the ruler and protractor axioms). We then define operations of addition and subtraction for line segments and angles which correspond closely to the addition and subtraction of arithmetic. Such a geometry is known as a metric geometry. Its interrelatedness with arithmetic is a specific example of the power of mathematical systems mentioned earlier.

In vector geometry we define the elements to be directed line segments, each with both a magnitude (length) and a direction. The operations must be defined then in terms of both magnitude and direction. Despite this complication, some complex theorems of geometry become extremely easy in vector geometry.

Transformational geometry focuses upon the ways geometric figures can be moved. Such a geometry is really a kind of dual system. On the one hand, the basic elemental geometric figures and the congruence, similarity, parallel, and perpendicular relations are maintained. But in addition three basic moves of geometric figures — translation, rotation, and reflection — become the basis for an additional set of elements. An equivalence relation is established for these elements, and the operation "followed by" is used to combine them. Despite this double-system complexity, translational geometrics are intuitively appealing. The emphasis on motion makes them suitable for introducing many geometric concepts to young children.

The Search for Structure

We still have not filled in the all-important details of mathematics, of course. Since it is these details that you will be teaching, it is the relationships and structure between them that is important to you. Filling in that structure is beyond the scope of this book, for it would require several additional volumes. Ultimately, searching for structure between the details of various areas of mathematics is why mathematics courses are required for prospective mathematics teachers. If we believe the cognitive psychologists, teachers who cannot reveal underlying structures to their students cannot teach mathematics meaningfully.

Perhaps in the course of your education, you lost the structural aspects of mathematics and were overwhelmed by the burden of details. This is a good time for you to take our overview of abstract mathematical systems and review their relationship to the mathematics content you will teach.

5-5: Study Module

For Further Investigation and Discussion

1. The study of concept learning gained much popularity with the recent work of Bruner. Nevertheless, the development of the idea has a long history. An early discussion of concept is found in chapter ten of John Dewey's, *How We Think*. Boston: D. C. Heath and Company, 1933. Read this chapter and compare Dewey's explanation of concept with recent ideas about concepts.

2. Ausubel has said, "If I had to reduce all of educational psychology to just one principle, I would say this: The most important single factor influencing learning is what the learner already knows. Ascertain this and teach him accordingly." (1:vi) Does this statement complement or contradict Piaget's experimental psychology? Explain your response. Is it compatible with the work of Gagné? With the work of Gestalt psychologists?

3. Are advance organizers antithetical to the idea of discovery learning? Are there situations in mathematics teaching where it would be preferable for the student *not* to know in advance how ideas were to ultimately be related? Should teachers be completely honest the first time around, or should they approach a topic through a series of "successive approximations"?

4. For an example of how Ausubel would organize an arithmetic and algebra program, read pages 85-96 of Chapter 4 of Ausubel, David P. and Floyd G. Robinson. *School Learning*. New York: Holt, Rinehart and Winston, 1969.

5. Reread the Bruner and Ausubel articles on discovery cited in Study Module 2-4. Both Bruner and Ausubel are cognitive psychologists. It is easy to see how they disagree in these two articles. How do they agree?

6. Read Brownell, W. A. "Meaning and Skill — Maintaining the Balance." *The Arithmetic Teacher* 3 (1956): 129-136. It is frequently argued that the over-emphasis on structure in new mathematics programs will result in children who cannot compute. How might Brownell answer these criticisms?

7. For another view of the implications of cognitive psychology for classroom practice read Bruner, Jerome. "Some Theorems on Instruction Illustrated with Reference to Mathematics." *Theories of Learning and Instruction*, 63rd Yearbook, Part I. National Society for the Study of Education. Chicago: The University of Chicago Press, 1964, pp. 306-335.

8. *The Mathematics Teacher* 55 (October, 1962) is filled with excellent articles on the nature of mathematics. In particular you will want to read "The Nature of Mathematics" by Mina Rees,

"The Changes Taking Place in Mathematics" by Irving Adler, and "Topology: Its Nature and Significance" by R. L. Wilder. Compare and contrast the ideas in module 5-4 with the ideas in these articles.

For Lesson Planning

We have used examples from algebra to suggest that the purposes of mathematics might be much clearer to students if we would rearrange the order in which some topics are usually presented. For example, looking at solutions of simple quadratic equations might indicate a reason for studying the simplification of radical expressions.

Study a textbook you might expect to use in teaching. Can you find a section where a particular technique is developed, but not used until a later section or chapter? Plan a series of two or three lessons which would reverse the sequence to show first how the technique might be applied, then develop it. Include a brief advanced organizer in the lesson where the technique is first developed. You will find it helpful to think in terms of behavioral objectives when planning these lessons.

For Microteaching

". . . the first precondition for meaningful learning is that it take place under the auspices of a meaningful learning set." (2:128) Ausubel indicates that if a student wants to simply memorize facts, the result of his learning must be rote or meaningless. The implication is that one function of the teacher should be to encourage students to want to understand the lesson in a meaningful sense. This understanding is best accomplished at the beginning of the lesson by means of a careful introduction or "set." Establishing a "set" at the beginning of a lesson requires several of the following activities:

1. An indication of what is to be learned in terms of standard names or terminology.
2. An indication of the purpose of relevance of the topic to be learned.
3. Relationships to previous knowledge, where possible.
4. A statement of expectations—what the learner should be able to do at the end of the lesson and the criteria against which he will be judged.
5. An advanced organizer which will allow the student to see the relative roles of the facts which will be emphasized in the lesson.

A teaching action which is complementary to establishing set, is the establishment of closure at the end of the lesson. Many of the same elements used in establishing set can be quickly repeated to summarize and close the lesson. A skillful teacher can establish closure at the end of a class period by asking a series of questions which will prompt students to provide the summary and organization of ideas they have learned. Although some teachers prefer to leave a few lessons open-ended so that students will continue to explore ideas outside the classroom, even these sequenced lessons must be closed eventually. Therefore, the ability to establish both set and closure is a skill every mathematics teacher should possess. Microteaching provides a good opportunity to practice these skills.

Outline a lesson on a mathematics topic of your choosing. Then plan in detail how you will establish set at the beginning of this lesson. Remember that you will have to catch the students' attention, so try to include some activity for them as quickly as possible. For example, giving an old familiar problem as a "one-minute quiz" is a good activity for prompting a discussion of the relationship of today's lesson to previous knowledge. (Too many teachers assume that establishing set requires a lecture. Beginning the class with a lecture may turn kids off and result in quite the opposite reaction than what you intended!)

Present the introduction to your lesson (but not the lesson itself) to four or five students in a microteaching situation. Have your presentation videotaped or audiotaped. At the end of the presentation, take some time to get informal evaluation from the students. You may want to ask them if they thought your presentation was a good way to start a lesson, if they would now be interested in continuing the lesson, etc.

Keep this student reaction in mind as you view your videotape or listen to your audiotape. Is your introduction clear and to the point, yet brief? If you have a videotape, look for the reactions of the students. Did you give them something to do, or are they just waiting for the lesson to begin?

If possible, replan your introduction and teach it to another group of students. You should also be ready to sequence all of the skills that you have practiced in microteaching into a full lesson. If this is possible, concentrate on developing good closure for your lesson as well.

For Related Research

For research on advance organizers, see Ausubel, David P. "The Use of Advance Organizers in the Learning and Retention of

Meaningful Verbal Material." *Journal of Educational Psychology* 51 (1960): 267-272. You may also want to read Ausubel, David P. and D. Fitzgerald. "Organizer, General Background, and Antecedent Learning Variables in Sequential Verbal Learning." *Journal of Educational Psychology* 53 (1962): 243-249.

The fundamental work in concept formation is Bruner, Jerome S., Jacqueline J. Goodnow, and George A. Austin. *A Study of Thinking*. New York: John Wiley & Sons, 1956. Pages 81-90 provide an introduction to the idea of selection strategies, and Chapter 8 provides a good overview of the details of the experimental work.

A study in concept formation in mathematics is reported in Shumway, Richard J. "Negative Instances and Mathematical Concept Formation: A Preliminary Study." *Journal for Research in Mathematics Education* 2 (1971): 488-494.

A study which suggests that concept attainment strategies vary by age is Tagatz, Glenn E., Jane A. Layman, and James R. Needham. "Information Processing of Third and Fourth Grade Children." *Contemporary Education* 42 (1970): 31-34. A study which suggests that negative instances become more important in concept formation with practice is Freibergs, V. and E. Tulving. "The Effect of Practices on Utilization of Information from Positive and Negative Instances in Concept Identification." *Canadian Journal of Psychology* 15 (1961): 101-106.

References to Part 5

1. Ausubel, David P. *Educational Psychology: A Cognitive View*. New York: Holt, Rinehart & Winston, 1968.
2. _____. "Facilitating Meaningful Verbal Learning in the Classroom." *The Arithmetic Teacher* 15 (1968): 126-132.
3. _____. "Some Psychological and Educational Limitations of Learning by Discovery." *The Arithmetic Teacher* 11 (1964): 290-302.
4. Bruner, Jerome, Jacqueline Goodnow, and G. A. Austin. *A Study of Thinking*. New York: John Wiley & Sons, 1956.
5. Sawyer, W. W. *Prelude to Mathematics*. Baltimore, Md.: Penguin Books, 1955.

Appendix

Table of *t*, for use in determining the significance of statistics

Example: When the *df* are 35 and $t = 2.03$, the .05 in column 3 means that 5 times in 100 trials a divergence as large as that obtained may be expected in the positive *and* negative directions under the null hypothesis.

Degrees of Freedom	Probability (P) 0.10	0.05	0.02	0.01
1	$t = 6.34$	$t = 12.71$	$t = 31.82$	$t = 63.66$
2	2.92	4.30	6.96	9.92
3	2.35	3.18	4.54	5.84
4	2.13	2.78	3.75	4.60
5	2.02	2.57	3.36	4.03
6	1.94	2.45	3.14	3.71
7	1.90	2.36	3.00	3.50
8	1.86	2.31	2.90	3.36
9	1.83	2.26	2.82	3.25
10	1.81	2.23	2.76	3.17
11	1.80	2.20	2.72	3.11
12	1.78	2.18	2.68	3.06
13	1.77	2.16	2.65	3.01
14	1.76	2.14	2.62	2.98
15	1.75	2.13	2.60	2.95
16	1.75	2.12	2.58	2.92
17	1.74	2.11	2.57	2.90
18	1.73	2.10	2.55	2.88
19	1.73	2.09	2.54	2.86
20	1.72	2.09	2.53	2.84
21	1.72	2.08	2.52	2.83
22	1.72	2.07	2.51	2.82
23	1.71	2.07	2.50	2.81
24	1.71	2.06	2.49	2.80
25	1.71	2.06	2.48	2.79
26	1.71	2.06	2.48	2.78
27	1.70	2.05	2.47	2.77
28	1.70	2.05	2.47	2.76
29	1.70	2.04	2.46	2.76
30	1.70	2.04	2.46	2.75
35	1.69	2.03	2.44	2.72
40	1.68	2.02	2.42	2.71
45	1.68	2.02	2.41	2.69
50	1.68	2.01	2.40	2.68
60	1.67	2.00	2.39	2.66
70	1.67	2.00	2.38	2.65
80	1.66	1.99	2.38	2.64
90	1.66	1.99	2.37	2.63
100	1.66	1.98	2.36	2.63
125	1.66	1.98	2.36	2.62
150	1.66	1.98	2.35	2.61
200	1.65	1.97	2.35	2.60
300	1.65	1.97	2.34	2.59
400	1.65	1.97	2.34	2.59
500	1.65	1.96	2.33	2.59
1000	1.65	1.96	2.33	2.58
∞	1.65	1.96	2.33	2.58

From Henry E. Garrett, *Statistics in Psychology and Education* (New York: David McKay Company, 1966). Reprinted with permission.